CW00432684

THE LIVES OF

SPIDERS

THE LIVES OF SPIDERS

A NATURAL HISTORY OF THE WORLD'S SPIDERS

Ximena Nelson

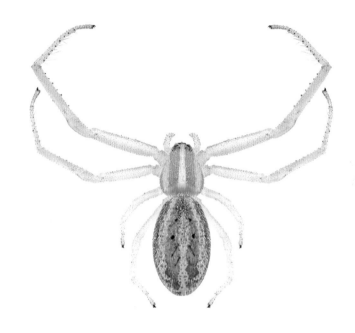

PRINCETON UNIVERSITY PRESS
PRINCETON AND OXFORD

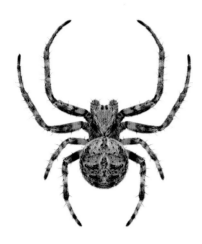

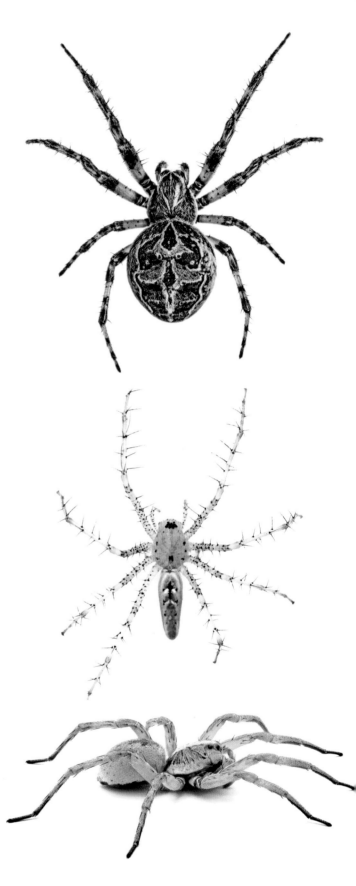

Published in 2024 by Princeton University Press
41 William Street, Princeton, New Jersey 08540
99 Banbury Road, Oxford OX2 6JX
press.princeton.edu

Copyright © 2024 by UniPress Books Limited
www.unipressbooks.com

All rights reserved. No part of this book may be reproduced
or transmitted in any form or by any means, electronic or
mechanical, including photocopying, recording, or by any
information storage-and-retrieval system, without written
permission from the copyright holder. Requests for permission
to reproduce material from this work should be sent to
permissions@press.princeton.edu

Library of Congress Control Number 2023943728
ISBN 978-0-691-25502-6
Ebook 978-0-691-25505-7

Typeset in Bembo and Futura

Printed and bound in Malaysia

10 9 8 7 6 5 4 3 2 1

British Library Cataloging-in-Publication Data is available

This book was conceived, designed, and produced by
UniPress Books Limited
Publisher: Nigel Browning
Commissioning editor: Kate Shanahan
Project manager: Kathleen Steeden
Designer & art directon: Wayne Blades
Picture researcher: Tom Broadbent
Proofreader: Steven Williamson
Illustrator: Sarah Skeate
Maps: Les Hunt

Cover images: (Front cover) fendercapture / Shutterstock;
(back cover and spine) alslutsky / Shutterstock

....

Dedicated to Robert Jackson, who instilled in me a lifelong
passion for spiders and the myriad of worlds in which they live.

CONTENTS

The world of spiders

Consider the feeling of the softest silk shirt against your shoulders, and ask yourself why a single strand of silk against your shoulders as you walk into a web should provoke anything but a pleasant reaction. Societally, we have an irrational fear of spiders, and we consciously or unconsciously teach that fear to our children, perpetuating the mistaken idea—and I do mean mistaken—that spiders are dangerous. An estimated 5 people in the global population of 8 billion people die of a spider bite each year. In contrast, about 750,000 people die of mosquito-borne diseases, and well over a million people die in car accidents annually, but we don't shriek "Car! Run away, kill it!" Part of the problem is that spiders are elusive—they are shy creatures, and most people don't know much about them. Those who *do* know about them realize that they are among the most diverse and interesting organisms on Earth. Welcome to the wonderful world of spiders!

UNIQUE LINEAGE

Spiders are not insects. The two lineages separated roughly 400,000,000 years ago—well before mammals came on the scene and, in fact, well before dinosaurs existed. Put another way, spiders have had a lot of time to evolve unique characteristics, which are plentiful. Spiders have some of the most sophisticated sensory systems known, remarkable predatory behaviors, and fussy eating habits (with one species being the world's pickiest predator); they can learn and perform among the most complex courtship behaviors of any animals.

Their world is one that humans cannot perceive, a world of tiny vibrations that tell the spider what it needs to know and—for web-building species (about half of all spiders)—one in which the web acts as an out-of-body nervous system. We could think of the web as externalized cognition—a useful trait when you have a tiny body and few neurons to process information.

↑ → One thing spiders have in common is that they are all predators, typically hunting insects as their main prey. However, even in their hunting tactics spiders differ: Some attack by stealth, some use webs to intercept prey, some ambush their target.

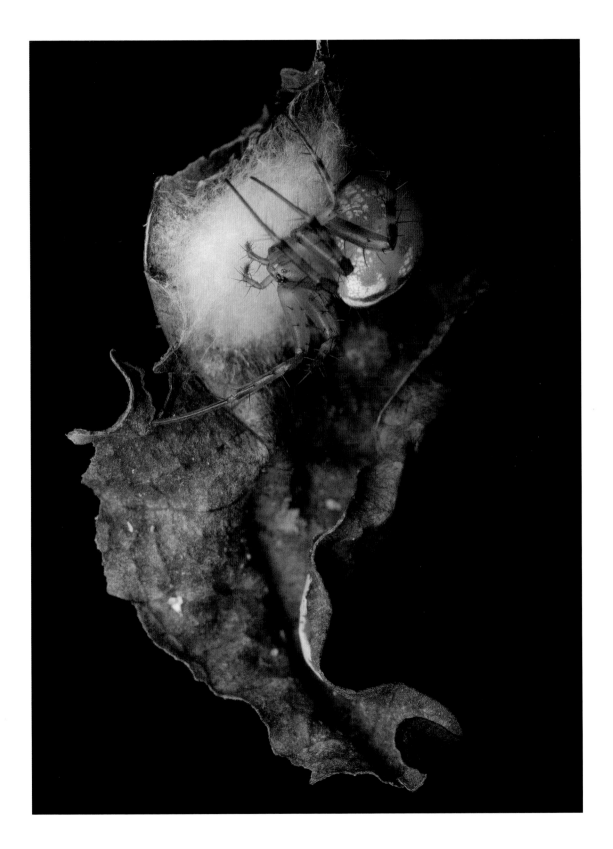

← Although insects are the main fare of most spiders, vertebrates are not immune to their hunting skills, and the occasional lizard, mouse, or bird may also provide a substantial meal for a hungry spider.

→ A spider's eight legs can be short and stocky (ideal for jumping and grasping) or, like this spitting spider, long and delicate (ideal for sensing vibrations beneath them).

While spiders are united in having two body parts (cephalothorax and abdomen) and 8 legs, the 132 family groups that currently make up the order Araneae (spiders) differ in almost every other respect, from their use of silk to their sensory abilities. Spiders vary enormously in body length and weight. "Spiderologists" (the technical term is *arachnologists*) do not include the legs when referring to spider size, and from their perspective, spider body length can vary from about $\frac{1}{100}$ in to 4 ¾ in (0.3 mm to 120 mm), corresponding to a 300 times difference in size, depending on species. Even within a single species, an adult female might outweigh the male by 125 times—a wild thought! This variation is apparent in all aspects of their lives: Some spiders live for less than a year, others can live more than 40 years. Some have four pairs of eyes, some have none.

ENGINEERING ROLE MODELS

Spiders can be considered nature's engineers. A major area of research is developing robots capable of interacting harmlessly with humans, because collision with traditional rigid robots is a bad idea. The applications of such robots include industrial

manufacturing and medical rehabilitation techniques. The use of soft materials for moving parts that are structurally sound and simultaneously elastic is integral to the development of so-called assistance robots. Because, unlike most other animals, spiders use a lightweight hydraulic system (rather than muscles) to straighten their legs, and combine this with immense power and sensitivity, the mechanics of spider locomotion has been used to develop soft actuators (the parts of a machine that convert energy into mechanical force, enabling movement). For example, the speed and pivotal flexibility of the hydraulically propelled jumps of jumping spiders are used to develop lightweight robots for jobs requiring high pressure and fast motion, such as for hydraulic pistons, and for safe-grasping tasks in robot–human interactions. More recent work is exploring spider-inspired soft robotics as prostheses.

Spider locomotion has inspired engineers to astronomical proportions, as depicted by the escape behavior of the desert-dwelling flicflac huntsman spider *Cebrennus rechenbergi*. This spider doubles its running speed by doing eight-legged cartwheels across the sand. The key is that the cartwheeling motion with many

legs maintains many points of contact with the substrate and has an even weight distribution—unlike, for example, bipedal motion, where one leg is raised off the ground with only one point of contact with the surface for periods of time. Based on this motion, engineers are developing articulated robots capable of navigating the rough terrain of other planets and the ocean floor. Spider-inspired space exploration does not end there, with labs building swarms of probes to make measurements of Mars' planetary atmosphere based on ballooning spiders.

SPIDER INSPIRATION

It is worth exploring how the knowledge of spiders has been used, or has the potential to be used, in science, technology, and medicine. Armed with a new understanding of spider sensory systems, materials, venom composition, and behavior, there has been a rapid expansion of bio-inspired engineering (biomimetics), biotechnology, and medicine based on spider biology. The range of spider-inspired technology is stunning and underappreciated: For example, the visual system of jumping spiders has led to miniaturized camera systems with exceptional depth of focus and optical resolution.

We have only begun to scratch the surface of how spiders can enrich our lives. Spiders also act as muses in music and the arts. They have inspired compositions for ballet, harpsicord, and percussion works, while in the visual arts, spiders have stoked the imagination in the creation of everything from paintings to sculptures to gigantic hanging web art installations.

In this book, we will explore the rich diversity of the lives of spiders, who unknowingly fascinate us and inspire a plethora of human endeavors.

← We think of spiders as being web-builders, but most do not build webs. These spiders often simply go undetected as they go about their business in the understory or in the high canopy. However, many web-builders build tiny, well-hidden webs, and they, too, go unseen—theirs is a life in which humans play no role.

WHAT IS
A SPIDER?

Spider anatomy and physiology

The physiology of a spider may seem simple at first glance, yet on closer inspection its complexities are nothing short of mind-boggling. The cephalothorax comprises a dual brain consisting of clusters of fine-tuned neurons that inform its delicately balanced nerves and sensory systems. The abdomen includes vital silk-producing organs, internal fluids that create hydraulic "muscle" movements, and the lower part of the digestive system.

Basic spider anatomy

The fused cephalon (head) and thorax (cephalothorax) of spiders differs from the separation of these body parts that is more typical of other arthropods. In spiders, the cepahlothorax and abdomen are joined by a narrow pedicel, which allows these body parts to move somewhat independently.

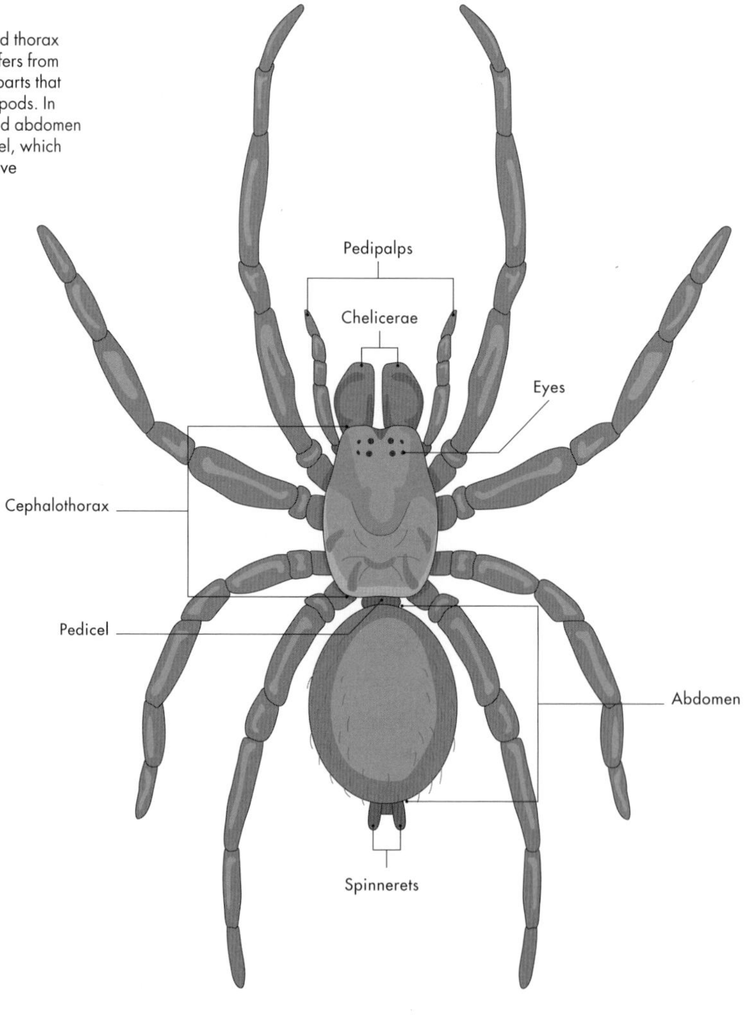

Pedipalps

Chelicerae

Eyes

Cephalothorax

Pedicel

Abdomen

Spinnerets

THE CEPHALOTHORAX

Housed within the cephalothorax of the spider is the central nervous system. This consists of two parts: the subesophageal and the supraesophageal ganglia. These ganglia are clusters of neurons, so we can imagine this as the spider having two brains. Each can be smaller than a poppy seed: one below and the other above the esophagus, although it is the supraesophageal ganglion that is often referred to as the "brain." Intriguingly, much like in humans, aging affects the spider's central nervous system, leading to a reduction in brain size and impaired behavior, such as reduced web-building ability and mobility.

The supraesophageal ganglion coordinates movement of the mouthparts, or chelicerae, and receives visual information from the multiple eyes of spiders, meshing the information to create a unified perspective of the surroundings. The subesophageal ganglion coordinates movement in response to information received by the sensory systems and has neural projections, or nerves, running to the venom glands, each of the four pairs of legs, and the pedipalps, or palps, which lie behind the chelicerae with their articulated fangs. The palps are often used to manipulate silk and prey items and, in males, are modified for sperm transfer to the female. All of these appendages are attached to the cephalothorax of the spider. Within the cephalothorax also lies some of the digestive system, which extends into the abdomen.

↑ The particular arrangement of the four tiny pairs of eyes designates this spider, waiting to ambush its prey, as a crab spider (Thomisidae).

THE ABDOMEN

In addition to containing the remainder of the digestive system, the abdomen contains the heart, the silk glands, and the respiratory, excretory, reproductive, and circulatory systems. In females, the ovaries are also in the abdomen, and female genital openings lie under the abdomen. Depending on the species, there is tremendous variation in the number of eggs a female can produce, ranging from one (in a single eggsac) to more than 3,000. At the rear of the abdomen, spiders have up to four pairs of spinnerets (usually three), which, as their name implies, deploy the silk with which many species spin their webs, build cocoons for their eggs, and build or line their safety retreats. Spiders use different types of silk for different functions and have up to eight different silk glands to produce these.

Spiders are very good at externalizing their bodily functions—and not just because the web can be considered an external sensory, or even cognitive, organ.

After injecting venom to paralyze and subdue their prey, spiders begin their somewhat unusual digestion, in that much of it occurs outside of the body. Known as extraoral digestion, spiders regurgitate digestive fluids containing enzymes that break down the internal soft tissue of their prey. Once the digestive fluid has acted, the spider sucks up and filters the remains of the digestive fluid along with the liquefied prey, often leaving behind the indigestible rigid exoskeleton.

Efficient conversion of food into energy (metabolism) is a requirement of an active lifestyle. As a group, spiders have unusually low metabolic rates, having reduced energetic needs through the use of venom to subdue prey and often relatively sedentary lifestyles. This is aided by reducing their need for metabolically costly and oxygen-hungry muscles, which are only used to flex limbs. Instead of having extensor muscles like almost all other animals, spiders shunt internal fluids into their joints to create forces (hydraulic forces) to straighten their limbs, resulting in rapid and often jerky leg motions. Nevertheless, they have to provide oxygen to their bodies in order to function.

Spiders breathe air and use multiple systems to do so. One or two pairs of "book" lungs, which evolved from gills, form the major part of the respiratory system, although some species have dispensed with book lungs altogether and use a tracheal respiratory system instead. With tracheal respiration, a network of transport tubes directly penetrates specific organs that require oxygen. Trachea diffuse gas at their tips, in direct contact with an organ, and also along their length. Most spiders combine diffusion-based book lungs, in which pores controlled by muscles open directly into the environment, with a tracheal respiratory system, which also has direct openings to the environment. However, the reliance on tracheal respiration is strongly correlated with activity level. Active spiders, such as jumping spiders, must meet the additional oxygen demands imposed by this lifestyle. Because tracheal systems are more effective at gas exchange, spiders with more active lifestyles that require better aerobic capacity have developed more complex tracheal systems.

Book lungs

Book lungs can be imagined as an open accordion, with the "bellows," or lamellae, being washed over by hemolymph (blood) for gas exchange, or respiration.

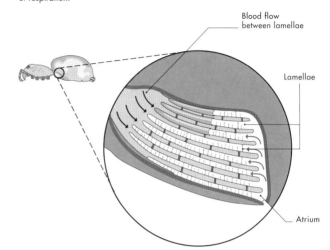

Blood flow between lamellae

Lamellae

Atrium

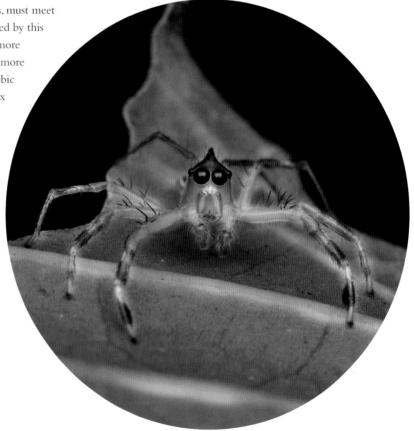

↖ Spiders, like this jumping spider, cover their prey in regurgitated digestive enzymes to begin the process of digestion outside their own body.

→ Jumping spiders are the most active of spiders, roaming the environment to spot and then pursue or stalk their prey before jumping on it. This requires good vision and an efficient mechanism to supply the internal organs with oxygen.

SENSORY FINESSE

The sensory systems of spiders are equally varied. Most spiders have four pairs of eyes, although some have three, two, one, or even no eyes. Wolf spiders and jumping spiders are highly visual and often use elaborate visual displays for communication. In terms of relative eye size, the exceptional vision of jumping spiders has no comparison among any animal. Although their eyes are typically smaller than $\frac{1}{50}$ in (0.5 mm), jumping spiders have visual spatial acuity, or visual resolution, comparable to many mammals and birds. Spatial acuity is what is checked during an eye test and when you read the bottom row of letters on an optometrist's chart. While there are animals, like us, that have better spatial acuity, they have eyes several times larger than an entire spider.

↖ Providing an indication of the importance of vision to their lifestyle, the eyes of jumping spiders can take up almost half of the space within the cephalothorax.

↗ Other spiders, like this wandering spider (Ctenidae) rely on other sensory systems, like mechanoreception (the ability to detect particle displacement, or vibrations), over vision to locate their prey.

All spiders also use pheromones, or chemicals secreted outside of the body, for communication. Spiders typically detect these pheromones by "taste," where they must physically make contact with the chemical (usually deposited on silk) and can sometimes detect them by "smell." Given how wide-ranging the use of taste, or "chemotactile" (and possibly odor-based) information is in spiders, we know remarkably little about the sensory organs that are involved in this process.

For most spiders, the most important sensory system is based on detecting tiny vibrations, either transmitted through the air (hearing), along the ground (so-called "seismic" information), or along the strands of silk that form the web. Using a complex sensory system comprised of the hairs on their bodies, spiders are exquisitely capable of detecting the smallest movement of particles. Some species are so sophisticated that they are believed to have the most sensitive sensory system of any type of any animal.

SPIDER KINEMATICS

Jumping, in which a rapid extension of the legs provides impulse for airborne locomotion, needs large forces to be produced quickly and synchronized accurately. Jumping spiders very accurately leap to catch prey and to move about, or, in a less directed manner, to escape from a predator. Prior to leaping, jumping spiders raise their first pair (sometimes the first two pairs) of legs, and attach a safety dragline to the ground in case things go awry. They propel themselves off the ground using the third pair of legs, or the third and fourth pairs, and seem to use all their legs to orient their body during flight. Among jumping arthropods, jumping spiders are neither the fastest, nor the longest-distance jumpers; nevertheless, these jumps can have take-off velocities of over 1.5 mph (2.4 km/h) and accelerations of 4–6 g (g-force). Although jumps are typically short, I have personally seen jumping spiders jump over 20 times their body length in apparently directed jumps (for example, both to attack prey and to move about), rather than in randomly directed escape jumps. So how do they do this?

Many arthropods, such as locusts, jump using catapult mechanisms, in which energy is stored and suddenly released. However, the few jumping spiders (adult weight ranging between 15 and 150 mg) that have been studied for the ability from which they take their name, suggest that they do not use this system. Instead, they probably rely on the slower system of muscle contraction (with the muscles actually being in the cephalothorax) combined with quickly shunting hemolymph into the legs, for further propulsive power, to jump. This is because spiders have a "semi-hydraulic" system of locomotion: Most animals have separate muscles used to flex and to extend the legs, but spiders only have flexor muscles. Leg extension is achieved because spiders contract muscles in the cephalothorax which generates a local increase in hemolymph pressure. This is conducted via the circulatory system to compartments in the legs and joints, increasing the hydraulic pressure to extend the legs.

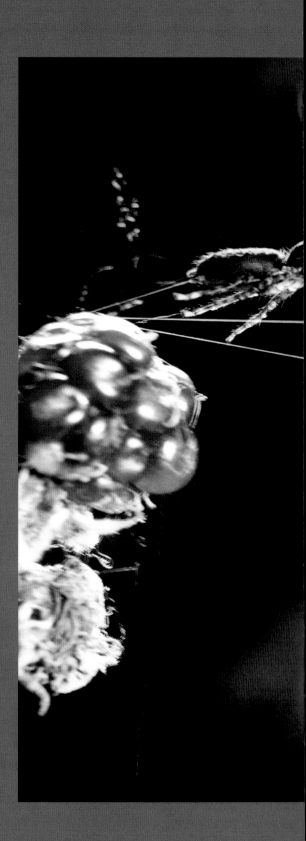

→   A jumping spider captured in action reveals not only how targeted their leaps are, but also that they attach a silk dragline as a safety line should things go wrong. This spider has jumped over four times its body length—in human terms, near 25 ft (8 m) from a standstill!

How to differentiate spiders

There are more than 50,000 species of spiders, with more being described almost every week. This makes spiders one of the world's most diverse animal groups, yet to date less than 15 spider species have been sequenced (about 0.02 percent of the total). With so little information to go on, it is hard to know why their genome is so heavily endowed. For most spider families, we currently don't even have easily identifiable genetic markers.

Extremely little is known about spider genomes because they can be exceedingly hard to sequence. However, ranging between about 1.5 and well over 10 gigabases, or Gb (1 Gb is 1,000,000,000 bases, the building blocks of the genetic code), they are huge. The human genome is about 6 Gb, and those of most insects are less than 1 Gb, which is why insect genomes have been studied extensively. It is unclear why spider genomes are so massive. This may be because there is a lot of redundancy within the genome, coupled with considerable coding for group diversity in their varied habitats, as well as in their physiological and morphological (pertaining to shape or form) adaptations.

← Spiders come in many shapes and sizes, from bulbous to skinny, and can be extremely colorful or dull brown.

↗ Scanning electron micrographs showing the unraveled embolus, the male copulatory organ (top), and some of the many silk-producing spigots on the spinnerets at the rear end of the spider (bottom).

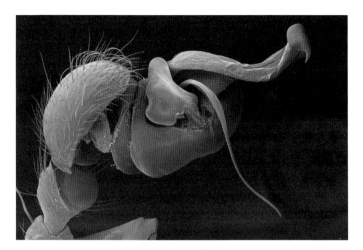

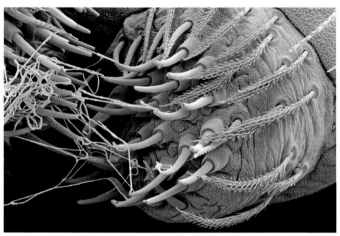

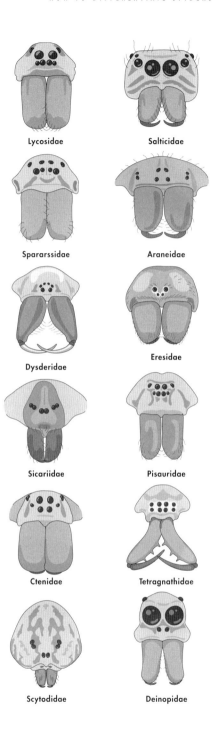

Lycosidae

Salticidae

Spararssidae

Araneidae

Dysderidae

Eresidae

Sicariidae

Pisauridae

Ctenidae

Tetragnathidae

Scytodidae

Deinopidae

MICROSCOPIC ANALYSIS

Another approach for identification, of course, is to look at a spider by eye or at a microscopic level. One of the key large-scale characters used to identify spiders into the major groups, or families, is the number, size, and positioning of the eyes, as well as the number and position of the spinnerets, and the claws at the end of their legs. However, the major characters used for identification to lower levels, such as species, are the reproductive organs, which are difficult to see without a microscope and usually absent in juvenile spiders, making juveniles particularly hard to sex or even to identify correctly.

Are you looking at me?

The positioning of the different pairs of eyes in spiders provides a clue as to the family to which they belong.

Male spiders are often, but not exclusively, smaller and more colorful than females and typically have a narrower abdomen. However, since the abdomen is very expandable, this is a poor way of sexing: It could just as easily be a hungry female. The most effective way to sex adult spiders is to take a close look at their palps: In adult males, since these are modified for reproduction, they have a swollen tip and are spoon-shaped, whereas female palps simply look like short, blunt legs.

Males vs. females

Male spiders are often smaller than the females. However, the key to sexing a spider is to look at the pedipalps, or palps, which in males are bulbous and somewhat spoon-shaped. This trait is usually only observable in sexually mature or near sexually mature males, making juvenile spiders essentially impossible to sex.

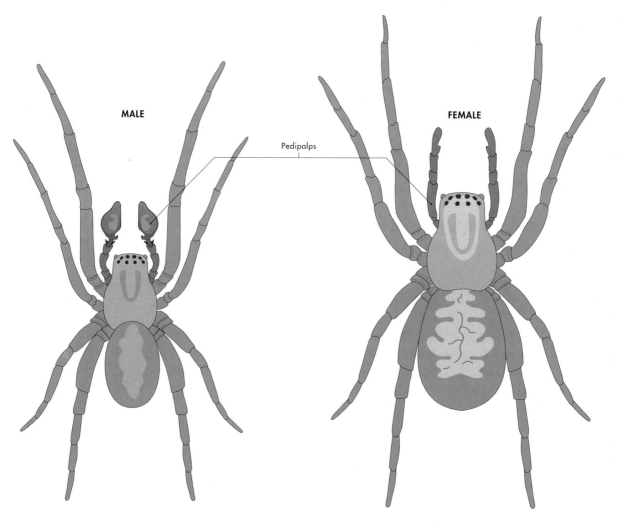

MALE

FEMALE

Pedipalps

ANCIENT SURVIVORS

The oldest group of spiders that still exists is the Mesothelae, of which there are only about 140 species in a single family, the Liphistiidae. These spiders have the ancestral mouthpart orientation in which the fangs on the chelicerae point downward. The remaining spiders are in a sister group that has two divisions: the mygalomorphs and the araneomorphs. Mygalomorph spiders are typically large and heavy-set burrowing spiders, with primitive downward-pointing fang orientation, while the fangs of araneomorphs point sideways and move in a scissorlike manner.

Why are spiders important?

Spiders are estimated to eat between 400 and 800 million tons of prey, consisting mainly of a wide variety of insects, every year. This alone means that they play a major role in food webs. Since they exist on all continents except Antarctica and live in diverse habitats, including forests, grasslands, wetlands, caves, deserts, and mountains, spiders have an impact on virtually all terrestrial ecosystems, and in many of these they are the main arthropod predators.

Because of this, spiders are believed to be good indicator species of the health of ecosystems. Habitat that is in poor condition will not sustain enough prey to support spider populations, so if spiders are faring poorly, this is a good indication that the habitat is degraded.

Many species of spiders share common habitat and compete for prey or space. Species that fall into interspecific competitive categories can be called "guilds," which in spiders are largely driven by the strategy devised to hunt prey. Hunting strategies include, how, when, and where the spiders hunt. Spiders differ on whether they are specialist predators (which tend not to compete with other species for the same prey; as such they have a niche diet, so technically do not fit the definition of a guild) or guilds consisting of generalist predators, to which most spiders belong.

Among generalist spiders, guilds can be identified by whether, for example, they are ambush hunters, like crab spiders, or active hunters, like wolf or jumping spiders.

ECOSYSTEM SHARING

Generalist spiders that use silk to catch prey can also be identified based on the type of web construction the species employs, as each web geometry will vary in how effective it is at catching different types of prey. These include sheet webs, space webs, and orb webs, for example (see page 72). The notion is that while species in each of these guilds will compete for shared resources, there will be limited overlap in competition for resources with other guilds of spiders. Essentially, this allows multiple guilds to share the same habitat, and each guild might have its own unique set of characteristics that may provide a valuable function to a given ecosystem or habitat. For example, web spiders are likely to regulate the levels of flying insects in a given area, while ground hunters are likely to exert predatory effects on ground-based walking insects, and each of these provides a function to the well-being of the entire ecosystem.

→   Crab spiders are proficient ambush hunters. "Loitering" among foliage, they are quick to attack even prey much larger than themselves (like this locust borer beetle) if they land or wander nearby.

Broad ecosystems comprise a variety of microhabitats, each with slightly different characteristics. For example, a shrubland ecosystem consists of multiple species of shrubs, each having different physical characteristics and flowers (which may attract different types of insects), as well as grasses, pebbles, soil, etc. The specific details of the structure of habitat are key drivers of spider distribution. Web spiders may be picky about the space required to build a web between shrubs, or ground-dwellers may be picky about the type of substrate (pebbles or soils) in which they burrow. Other, more active hunters that build retreats and actively search for prey away from them may prefer to build their retreats on leaves of a specific size and texture, thus limiting the types of plants in which they will be found. This means that there are numerous microhabitat choices being made by spiders that,

in turn, allow them to cohabit within the same habitat and reduce competition for food or locations in which to build a safe retreat. Another way of reducing intraguild competition, of course, is to live in a place where you have extremely limited or no competition from other spider species, such as underwater.

↑→ Every level of a habitat will contain spiders: They will be present burrowing under the soil, wandering on the forest floor, residing in the the low-lying bushes, and using the limbs of trees to make webs high in the canopy.

SURPRISING HABITATS

Several unusual habitats are used by spiders, with caves
and underwater possibly being the oddest. These habitats
also require specific mechanisms to enable spiders to live
there. The diving bell spider, *Argyroneta aquatica*, spends
most of its life underwater, where it hunts aquatic
invertebrates. For an air-breathing terrestrial invertebrate,
clearly some changes have been made that have allowed
this spider to successfully exploit this new habitat.
Underwater, it builds a web between aquatic plants,
much like its relations do above water, and attaches
a large bubble, or "diving bell," to the silk. This bubble
can be large enough to house the entire spider, which
can breathe within the bubble; or if it is smaller, it can

↖ Using hydrophobic hairs,
spiders can trap a bubble of air under
the external openings to the respiratory
system to provide air on short forays
underwater.

↑ The diving bell spider (*Argyroneta
aquatica*) carries the air bubbles down
from the surface to the silken "diving
bell," forming an air pocket for the
spider to live in. It will make occasional
trips to the surface to replenish the air
with new air bubbles.

↗ Spiders that live in caves, like
this Barr's cave spider (*Nesticus barri*),
have often lost the use of vision due to
the limited use of this sensory system
in areas with little to no light.

accommodate the abdomen, which houses the external openings to the respiratory system. The spider replenishes oxygen in the diving bell by going to the surface to collect a bubble of air, which it attaches to its abdomen and takes down to refill its supplies.

Cave spiders have a different set of adaptations to cope with the difficulties posed by the environment in which they live. Deep in the bowels of the earth, cave-dwelling species, known by the lovely word *troglobionts*, live in areas in which there is little to no light. Consequently, they have little use for vision, and some spiders, such as *Sinopoda scurion*, have lost all eyes, while several newly described species of *Tegenaria* funnel web spiders in caves in Israel are in various stages of eye reduction, depending on how deep within the caves they live. Having limited prey available to them, troglobiont spiders also appear to have very slow development and an especially low metabolic rate. In addition to many species that are associated with caves (such as those found at cave entrances), more than 1,000 spider species from 46 families are known to be troglobionts. Unfortunately, because of where they live, little is known about them, highlighting not just the diversity of spiders and their role in countless ecosystems, but how poorly understood these animals are.

Spiders through time

Fossils contain information on when, where, and what organisms looked like throughout geological time, but are only useful when they are available. However, the fossil record is patchy and is biased against soft-bodied organisms, which don't preserve well. This includes spiders. Consequently, spider fossils are relatively scarce unless they are found in amber, fossilized tree resin.

Because of its very nature, amber is also biased, in this case toward spiders that live and hunt on tree trunks. Some of these amber-preserved specimens are very clear, and eye placement or spiders trapped with prey items can even be discerned. However, amber-preserved spiders only became abundant about 66 million years ago (mya), and rare fossils show that the spider lineage is considerably older than that. Therefore, until recently, a clear picture on early spider phylogeny (the relationships between the groups of organisms, or the evolutionary tree) was murky. However, advances in molecular DNA sequencing techniques on extant spiders (spiders present at this time) have now allowed a clearer

understanding of the evolutionary history of this group. Phylogenomics relies on molecular techniques to get a better understanding of divergence times (when they separated) between groups at all levels, from high-level taxonomic groups, such as classes and orders, to fine-grained lower-level classifications, such as family, genus, and even species.

The subphylum Chelicerata lies within the phylum Arthropoda, which also contains the separate subphylum Hexapoda, to which insects belong. The class Arachnida, which contains mites and ticks (superorders Acariformes and Parasitiformes), scorpions (order Scorpiones), harvestmen (Opiliones), camel spiders (Solifugae), false scorpions (Pseudoscorpiones), Ricinulei (no common name), micro-whipscorpions (Palpigradi), whip scorpions (Uropygi), short-tailed whip scorpions (Schizomida), tailless whip scorpions (Amblypygi), and spiders (Araneae), is within the Chelicerata.

↘ The evolutionary history of spiders depicts how spiders arose within the larger chelicerate group, itself a part of the older arthropod group which also contains insects (Hexapoda).

535 MILLION YEARS AGO	500 MILLION YEARS AGO	420 MILLION YEARS AGO	400 MILLION YEARS AGO	380 MILLION YEARS AGO	270 MILLION YEARS AGO
First Arthropoda	First Chelicerata	First arachnid (class Arachnida)	First insects (class Hexapoda)	First true spiders	Spiders radiate widely

PALEOZOIC

Timeline not to scale

ANCIENT ORIGINS

The fossil record suggests that terrestrial colonization of the Chelicerata from their marine ancestors occurred about 440 mya in the Ordovician period. The oldest "almost spider," which has traits in common with spiders, tailless whip scorpions, and whip scorpions, dates to more than 370 mya, but the earliest true spiders can be assigned to about 300 mya. Spiders really came into their own and began diversifying on a grand scale about 100 mya. Today, it is believed that about half of the spider species remain unknown or undescribed to science, so estimates suggest that there are currently about 100,000 species of spiders in 132 known families. To put this into context, the earliest mammals evolved within the last 200 million years, and although they diversified widely after the extinction of the dinosaurs about 66 million years ago, there are less than 6,500 species of mammals. This is about the same number of species currently found (bearing in mind that a good percentage remain undescribed) in the largest single family of spiders, the jumping spiders.

↗ The "living fossil." Ancient *Eriauchenus*, pelican spiders, specialize in hunting other spiders, which they can attack from a safe distance as they are able to raise their chelicerae about 90 degrees.

→ Spiders caught in amber can be beautifully preserved, sometimes even found with their prey, providing an insight into their behavior tens of millions of years ago.

| 228 MILLION YEARS AGO First dinosaur fossil | 150 MILLION YEARS AGO First bird fossil | 65 MILLION YEARS AGO First primate fossil | 6 MILLION YEARS AGO First hominim | 2 MILLION YEARS AGO First humans from the genus *Homo* |

| MESOZOIC | CENOZOIC |

SPIDER
SENSES

Sensory systems

Like other arthropods, many of the sensory abilities of spiders, which include the ability to detect or determine humidity and temperature, as well as more familiar senses, such as taste and touch, rely on the hairlike sensilla that cover their bodies. Sensory organs within multiple types of sensilla mediate responses to different forms of stimulation, such as heat rather than touch. In addition, some spiders rely on olfaction and vision. With advanced technology, we are only now beginning to uncover the wonders of spider sensory systems.

No matter what the lifestyle or size of a spider, it still needs to have a very clear sense of "self" in terms of its movement and the location of its many limbs. Like us, spiders need to ensure that all limbs are where they should be and, for instance, not crashing into things. For this, spiders rely on a diversity of very detailed sense organs, including sensilla, to measure internal and external forces on their leg joints. This fundamental yet often overlooked ability is called proprioception and relies on mechanoreceptor cells that have the ability to respond to mechanical stimulation, such as bending or pressure, to know where the tip of the leg might be,

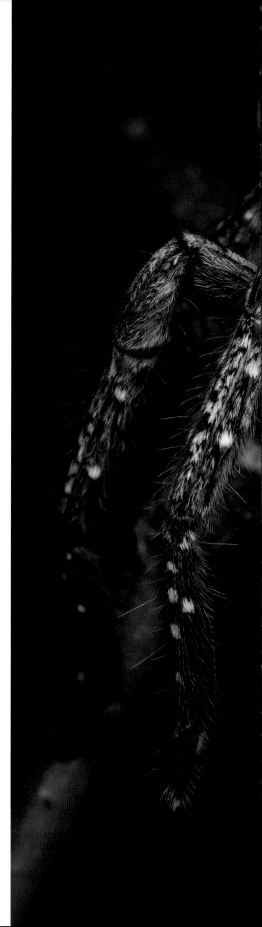

→   The hairy legs and bodies of spiders, like this huntsman, serve an important purpose: By deflecting to the tiniest movements of air around them, each hair tells the spider about the world surrounding it.

given even the tiniest of movements or flexions of one of
a segment of one of the eight legs. All this requires sophisticated
neural processing of sensory information, and when it comes to
mechanoreception, spiders have few rivals among any animals.
However, spiders are not one-trick ponies and could be described
as jacks-of-all-trades and masters of all. Once again, this is reflected
in the needs imposed by the diversity of their lives.

DIVERGENT SENSORY NEEDS

A spider walking through leaf litter searching for prey has different
sensory needs to a spider hanging in its web. The ground spider
might rely more heavily on vision than its web-building
counterpart, which in turn might rely more on vibrations. Similarly,
nocturnal spiders are less likely to rely on vision than diurnal

→   What may appear as a tangled
mess of silken strands is often
deliberately constructed to provide the
best possible vibrational information in
what can be considered a sensory
system external to the spider's body.

↓   Leaf litter is not very good
at transmitting vibrations of prey
wandering nearby, so many spiders
hunting in this habitat rely on
ambushing prey that haplessly
wander past.

spiders. These are certainly not strict rules, but there is a correlation between the foraging lifestyle of the spider and both its reliance on a given sensory system and its use of that sensory modality for communicating.

In addition to their proprioceptive abilities, spiders use mechanoreception in some form or another to obtain information about the world and to communicate with each other. For example, spiders communicate using their acutely sensitive ability to respond to pressure during tactile or touch-based communication and also through "sound"—their ability to detect particle vibrations though a given medium, such as the air or ground. Spiders also rely heavily on visual and chemical sensory modalities.

With some exceptions, most notably vision in jumping spiders, we still know relatively little about the sensory systems of spiders. Recent work suggests that spiders may "hear" sound waves propagating through air, as we do. In 2022, it was discovered that a web can act as an acoustic antenna that amplifies the distance at which sound can be detected. While spiders have so far amazed scientists, artists, and engineers, there is no doubt that their sensory systems may yet hold many more surprises for us.

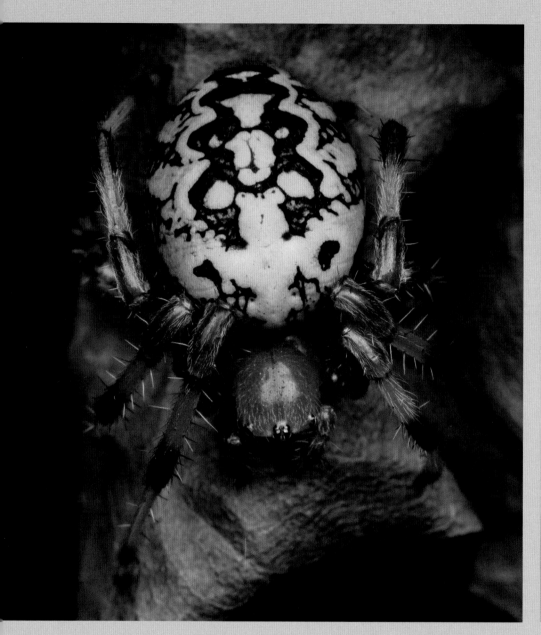

↑ A spider on a leaf has to contend with differentiating vibrations across the surface of the leaf, and discriminating between leaf movement due to wind or a potential predator, such as a bird.

↗ Spiders can be very cryptically colored on bark, with a flat body, making some trunk-dwelling, ambush-hunting spiders near impossible to detect. The spider, however, will notice any movement of air, or possibly prey odor, as it sits still and waits.

Chemoreception

The ability to detect information from chemicals is known as chemoreception and is often performed via two channels: Olfaction discerns airborne chemicals, while chemotactile information, or taste, requires physical contact between the sensing organ and the chemical in question.

Chemoreception is thought to be the most ancient of the sensory abilities in animals, yet, paradoxically, it is one of the least understood. This is especially true for spiders, where behavioral evidence for the use of olfaction and chemotactile information abounds, but where an understanding of the mechanisms used to

detect chemical information, especially airborne chemicals, are sorely lacking. In fact, we are still unsure about where many of the chemical sensory receptors, or chemoreceptors, actually are.

Spiders use chemical information in one of two ways. The first is straightforward signaling for communication, where the recipient of the signal adaptively changes their behavior upon detecting the signal of a member of the same species, and this was the intended outcome from the sender's point of view. Chemical signals like this are known as pheromones and often include sex-based information informing the recipient of the sex and age of the emitting spider.

However, spiders can also adaptively change their behavior upon detecting a chemical that was not intended to do so by the emitter, which is often not a member of their own species. This can be thought of as eavesdropping on chemical information. These are called cues and include, for example, the ability to detect the chemical cues that might give away the presence of nearby potential prey. From the prey's point of view, this is not a signal; it was evidently not intended to elicit searching behavior in the receiver, but being able to detect this is highly advantageous from the point of view of the receiver, which may end up nicely engorged with food.

↑ *Scytodes* spitting spiders (see page 128) are nearly blind but seem to have an uncanny ability to detect the presence of prey or danger through smell and "taste" (contact chemoreception). On detecting potential prey—typically other spiders—they expel silk to snare the potentially deadly prey from a distance.

← Wandering spiders (Ctenidae) typically have more mechanoreceptive hairs than web spiders, suggesting that detecting airflow and ground-based vibrations are particularly important to them, but they also have chemoreceptive hairs, allowing them to perceive chemical compounds of interest.

Chemoreceptors in sensilla

Spiders have curved, blunt hairs concentrated on the tips of their legs and palps that are used for taste and sometimes smell. These are also called "tip-pore sensilla," because they have a pore open to the environment at their tip, which leads to a central canal within the shaft of the hair, or sensillum.

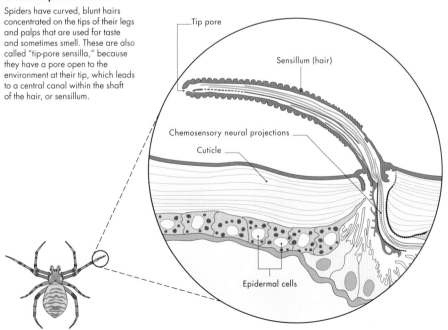

Tip pore

Sensillum (hair)

Chemosensory neural projections

Cuticle

Epidermal cells

SIGNALS IN SCENT

All spiders seem to respond to chemotactile pheromones, but the use of olfaction seems to be more limited. Chemotactile signals are often deposited on silk, which is almost universally used by spiders. Web-builders use silk for construction of the web (among other uses), while burrowing spiders often line their burrows with silk and may leave silken trip lines extending out of the burrow, alerting the spider inside of any potentially edible thing that might walk over the silk near their burrow. Finally, spiders that do not build webs, but instead wander around the environment (wandering spiders) typically deposit walking draglines as safety anchors as they walk. Silk is therefore a good indicator of the presence of a spider, and if signals are deposited on that silk, another spider might be able to glean important information about its owner.

Behavioral evidence suggests that spiders can obtain a level of information from chemotactile signals and cues that are difficult for humans to comprehend. For example, spiders can determine the size and hunger level of another spider. It might seem odd to be able to figure out if a nearby spider is hungry, but in spiders, which are intraguild predators of each other, it pays to know the level of threat posed by a nearby potential predator, even if that same individual may also be a potential mate. More mundane chemotactile information, such as whether a spider is a member of their own species, and the sex, virgin-mated status, and maturity level (an indicator of whether it is sexually mature or about to become sexually mature) of the spider seems to be commonplace among spiders of many types. For males in search of females, this is important information: They prefer virgin females, in part because this increases their chances of paternity and in part because they might not survive the interaction for another try. However, females, especially sedentary females such as web-builders, also face the problem that they may remain undetected and therefore unmated, and they increase their effort to lure a mate by increasing pheromone production as time runs out for them.

← Funnel web spiders build a dense silk-lined burrow, here in heather, which can be over 8 in (20 cm) deep. They will hide at the bottom until they detect the presence of something worth emerging for.

→ If you reach out to a spider, and it doesn't run away immediately, it will likely also reach out toward you with a hairy foot—all the better to "smell" you with.

DEFENSIVE CLUES

Spiders also use chemotactile cues from other species of spiders to determine threats, in addition to cues from members of their own species to determine the presence and size or condition of rivals. Yet, despite all of this behavioral evidence, details of the compounds that are used for signaling, and how and where these are detected, remain largely elusive.

Nevertheless, progress is being made. The chemical compounds of some spider pheromones, often comprised of acids and lipids, have been identified, and from the evidence available, they seem to be more diverse than those of insects. Although we still don't know where exactly the chemosensory structures that mediate olfaction are, there are thousands of chemotactile sensilla on the palps and tips of the legs. This might explain why spiders often daintily touch items with their appendages before walking on them. Try having a spider walk on your hand: It will reach out a leg and touch you before most likely turning away, which suggests that you don't taste too good! These same organs may also detect smell, which explains why many spiders will stop their activity and wave their first or second pair of legs around, possibly smelling something of interest in the environment, as a snake or a lizard does when it flicks its tongue out. Perhaps the spider approaching your hand raised its legs toward you but didn't even touch you—in which case you probably didn't smell too good either.

Mechanoreception

One of the first words that pops into people's minds when describing spiders is "hairy." They are indeed hairy, and many of these hairs are used to interpret the world around them in dizzying detail, simply by the fact that they can bend—or if they are stiff hairs, be deflected in their entirety.

Hairy feet

Spiders have sensory feet that are capable of detecting the slightest touch, puff of air, or even smell, depending on the type of hair responding to the stimulus.

In spiders, stimuli that can cause a mechanical change, such as deflection, bending, or pressure, are detected by a range of organs, including tactile hairs, slit sense organs, and articulated stiff hairs known as trichobothria. Mechanoreception, largely mediated by hairs, is the sense most widely used by spiders, and the range of sensory organs employed gives an indication of the importance of mechanoreception to survival.

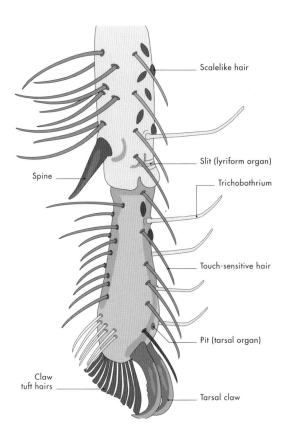

- Scalelike hair
- Slit (lyriform organ)
- Spine
- Trichobothrium
- Touch-sensitive hair
- Pit (tarsal organ)
- Claw tuft hairs
- Tarsal claw

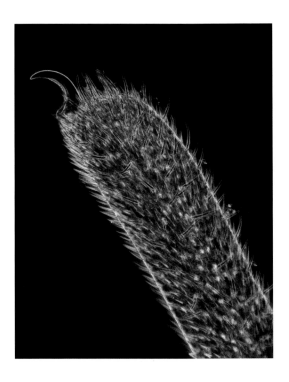

GOOD VIBRATIONS

Let's start with web-builders. Sitting either at the hub of the web or in a retreat attached to the web, these spiders rely not on vision but on vibrations to detect their surroundings. The thin strands of silk making up the web vibrate with the wind and shake if the web is struck. The shaking of the interconnected strands of silk, coupled with the geometry of these intersections, informs the spider about where the web has been intercepted. The amplitude, or strength, and frequency of the vibrations tell the spider about the size and strength of the struggling prey. As the prey tires from its struggles, this, too, will be transmitted through the silk as a change in amplitude and frequency in the vibrations, thus telling the spider about when it is safe to approach. Similar information can be obtained by burrowing spiders, such as some trapdoor spiders (which have an openable lid covering the burrow entrance), which extend radiating strands of silk from within their burrows (where they reside safely out of

view from predators) to the outside. Here, any insect walking by the undetected burrow will trigger vibrations in the given silken trip line it has walked over, thus telling the spider within not only that something is out there, but because of the individual strand of silk along which most of the vibrations are humming, also its location. Once again, the strength of those vibrations may tell the spider, much like the web-builder, how large and potentially dangerous the animal outside is. All of this information is detected by slit sensilla and mechanosensory hairs that bend and deflect with the vibrations passing along the silk.

↑   Silken strands of vibration sensors extend like Medusa's hair from a trapdoor spider's burrow.

←   Up close, we can see that a spider's leg has one or more tarsal claws, with which they can grasp onto silk strands, and is exceedingly hairy. Many of these hairs provide the spider with sensory information.

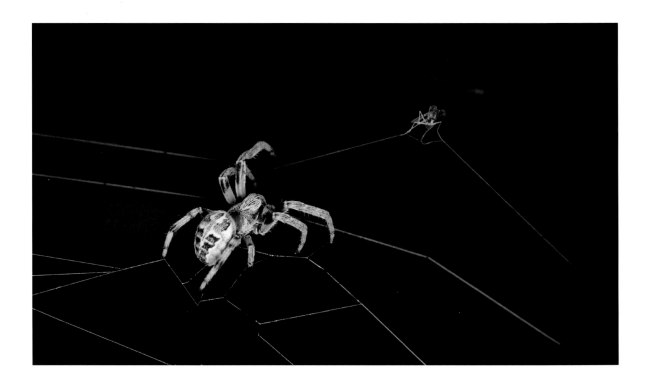

Movement that causes hairs to respond to vibrations along silk is also present in other media, such as air. Hearing requires hairs sensitive enough to bend or deflect with the movement of air particles that are compressed by sound, creating waves of compressed and decompressed molecules of air. For a given tone, there will be a characteristic wavelength between the peaks of each cycle of compression and decompression of particles. Hairs of different length or stiffness respond (for instance, deflect) best to specific wavelength frequencies, as in our inner ear: This is the basis of our tonal hearing—a high-frequency shriek is discernible from a low-frequency rumble. Recent work suggests that without ears, but instead using mechanosensory hairs, spiders also detect airborne sound, either directly, as in jumping spiders and ogre-faced spiders that can discern prey and predators 6 ½–10 ft (2–3 m) away through specific sound frequencies (tones) associated with their flight behavior, or in the case of web-builders, once again though the silk. In this case, the silk is so sensitive that the web itself pulses with sound at a distance and is thus detectable to the spiders, essentially transforming the web from a direct vibration sensor responsive to items contained within it, to a large antenna capable of detecting and localizing sound produced by a potential predator or prey from at least 33 ft (10 m) away.

GROUNDING ENERGIES

Yet another medium capable of transmitting vibrations is the ground. Anyone who has felt an earthquake will be well aware of this. Earthquakes contain a phenomenal amount of energy, while the pitter-patter of another spider or beetle walking about is another matter altogether, and these require very sensitive organs to be detected. Yet, organs sensitive enough to detect these tiny movements can also trigger appropriate responses in their owner. It is not as though smaller organisms somehow have better sensibility due to being small: This is constrained by biological needs, chemistry, and physics, and does not scale according to size. In fact, our understanding of some physics is challenged by spider sensory systems. Known as seismic vibrations, these substrate-borne movements can lead wandering spiders to potential mates, prey, or even rivals. Specific "drumbeats" can also be used as signals:

for example, by courting wolf spider males trying to impress females with the virility of their beat.

There are slightly different forces involved in vibrations transmitted through the air or the substrate, and consequently spiders use different sensory organs to detect them. Slit sense organs are found in some configuration (sometimes in groups, sometimes singly) throughout the body. These are tiny slits in the exoskeleton that respond to mechanical stimulation, due to the slight deformation of the cuticle that occurs when there is a seismic vibration or when there is strain in the exoskeleton during movement. Groups of these (known as lyriform organs) are often found on the leg joints, suggesting that they have an important proprioceptive role.

FLEXIBLE HAIRS

Bendable tactile hairs, of which spiders have tens of thousands, are hollow and respond to both very small and very large forces without breaking. These hairs respond to direct pressure, and so are crucial to proprioception, but also to direct vibrations, such as those passed along a strand of silk. In contrast, the narrow, stiff trichobothria are very sensitive to airflow, like that caused by a flying predator such as a wasp, or a potential prey such as a fly. Much like the arrangement of the hairs in our inner ear that give us our ability to distinguish between tones, the spider's tactile hairs often occur in rows, with each hair in a given clump being of a particular length. Because of the properties of fluid mechanics, beyond the scope of this book, long trichobothria respond well to airflow of low frequency, while short trichobothria respond best to high frequency, with mid-length hairs acting accordingly. Consequently, by having hairs of varying length, the spider can clearly discern a wide range of frequencies, some of which will indicate danger and others nearby prey. Using high-powered microscopy, we can see that in some species, such as *Cupiennius salei* (see page 56), these hairs have tiny filaments poking out of them. These dramatically extend the range of frequencies and sensitivity of the trichobothria, arguably making them the most sensitive organ so far discovered.

↑ A net-casting or ogre-faced spider can detect sound from prey flying nearby and rapidly deploy its expandable leg-held web to snare the prey.

↖ The spider will interpret the movements caused by this fly struggling on the web through the different types and energy characteristics of the vibrations down each stand of silk. This helps the spiders to determine the fly's location, its size, and how tired it is.

Vision

Unlike the compound eyes of insects, which look like a mirror ball of individual lenses, spider eyes have a single lens, like our own, and are known as simple or camera eyes. The camera eye lens faces the environment. Light entering the lens gets channeled onto the retina, which contains the cells that respond to light, the photoreceptors.

Photoreceptors are neural cells, and contain light-sensitive rhabdomeres that absorb light from the environment and provide the brain with visual information on shape, contrast, color, and motion. Unlike our eyes, which can swivel independently of our head to "face" the object of interest, spider lenses are part of the exoskeleton and are unable to move.

Spider eyes come in two flavors: principal and secondary. The principal eyes are formally called the anterior median eyes because they are often in the front and middle of the face. When spider lineages lose a pair of eyes, the principal eyes are usually the ones that are lost. The remaining three pairs, also named because of their relative position around the head (anterior lateral, posterior lateral, and posterior median eyes), are the secondary eyes. A number of things differentiate the principal and secondary eyes, and this starts right at the beginning: They arise from developmentally distinct tissue, leading to large structural differences, such as the secondary eyes being "inside out."

In the secondary eyes, the photoreceptors are positioned with the dense—and optically imperfect—cell bodies "facing" the light, while the long light-collecting portion of the cell (containing the rhabdomeres) is deeper inside the head, farther from the light entering the eye. This counterintuitive arrangement is not uncommon: It is the same situation with our own eyes. The principal eyes of spiders are the right way around, and because there is less scatter of photons of light due to intervening cell bodies (as in the secondary eyes), for optical reasons this is generally a better arrangement. Furthermore, the photoreceptors in the principal eyes are typically small and much more densely packed than in secondary eyes, allowing the principal eyes to see well, with excellent resolution or spatial acuity. In most cases, the principal eyes also mediate color vision.

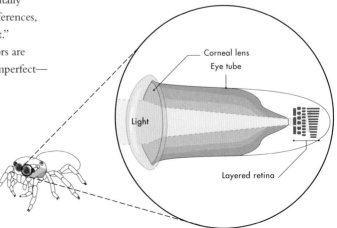

Principal eye

Morphology of the principal eyes of jumping spiders, depicting a long eye tube that acts as a telescope, at the end of which sits a retina with layers of photoreceptors.

Corneal lens
Eye tube
Light
Layered retina

↑   Most spiders have four pairs of "camera-type" eyes, and their arrangement or positioning on the head is a key indicator of the spider's taxonomic grouping, while the size and number of the eyes gives away the importance of vision to each spider group.

ULTRAVIOLET SIGHT

Many spiders can see in color, but it is a different color spectrum to our own. Color vision is possible because different photoreceptor types are sensitive, or respond, to different wavelengths of light, which correspond to different colors. Central nervous system processing of the information provided by these different photoreceptor types compares these responses, much like mixing primary paint colors to produce a perceived output color spectrum. In a hypothetical example, if you get equal responses of photoreceptors sensitive to yellow as to blue, you would perceive the color as green. Our vision is typically trichromatic, in that we have three classes of photoreceptors sensitive to three wavelengths of light, corresponding roughly to blue, green, and red. Most mammals are dichromats and can only compare the outputs of two classes of photoreceptors, severely restricting the color palette that they are able to see. Spiders that see in color exhibit the variation we come to expect: They can be dichromatic, trichromatic, and even tetrachromatic. However, one of the colors they are all able to see is ultraviolet (UV), which humans cannot.

REFLECTIONS ON TAPETA

With some exceptions, the secondary eyes do not support color vision, but can be especially good for detecting motion, contrast, and objects in low light. The secondary eyes of many families have a tapetum, a thin layer lying behind the retina that reflects light like a mirror back into the eye for the rhabdomeres to have a second chance at trying to extract photons from the available light. This is an excellent way of spotting nocturnal spiders: If you go out at night with a flashlight, and tiny globes at ground level reflect back at you, they are most likely wolf spiders and the light shining back is reflecting off their tapeta. Other families, such as net-casting or ogre-faced spiders and jumping spiders, do not have tapeta, possibly because the reflection of the light back into the eyes causes some optical problems when the aim is to provide the best possible spatial resolution.

RETINAL DIFFERENCES

Another major difference between the principal and secondary eyes is that the retinae of the secondary eyes is fixed—they can't move—but in many families the retinae (not the lenses) of the principal eyes are movable. This affords those spiders the ability to turn their gaze on something without giving away their presence by having to move their cephalothorax or body. These retinal displacements are best developed among jumping spiders, where a series of muscles enable horizontal, vertical, and torsional movements of the retinae, which are used to track and carefully assess objects of interest within the field of view afforded by the movements of the retinae. Retinal movement of the principal eyes is especially important to spiders that rely heavily on vision, because it is these eyes that are typically responsible for high resolution and color vision.

What this shows is that the two types of eyes work together, each performing their own specific and dedicated functions. In general, the secondary eyes provide the spider with motion-based information across a wide peripheral view, which tells the spider that there may be a mate, food, or danger nearby. This information is than used to guide the spider to orient its body, or if possible, its principal eye gaze (via retinal movements) toward the source, allowing these eyes to investigate the object in glorious detail.

← The large forward-facing eyes of jumping spiders indicate that they are visual predators. Compare this, for example, to grazing animals: Their eyes are positioned to the sides so that they can better see around them and escape their carnivorous predators, which in turn have forward-facing eyes to stalk and track prey.

LATRODECTUS HESPERUS

Western black widow spider

Web womanizer

SCIENTIFIC NAME	*Latrodectus hesperus*
FAMILY	Theridiidae
BODY LENGTH	Females ½–⅗ in (12–16 mm), males ¼ in (6 mm)
NOTABLE ANATOMY	Red hourglass shape on shiny black body in females
MEMORABLE FEATURE	Potentially venomous bite to humans

Much feared by humans due to their neurotoxic venom, "widow" spiders in the genus *Latrodectus* use chemical information, or pheromones, to glean crucial insight into potential mates. Like most spiders, the tiny *Latrodectus hesperus* males seek out their nine-times-larger female counterparts for reproduction. The male can discern between species based on odor; and based on contact with the silk-borne female pheromones present on a female's web, they can distinguish between females of different ages, hunger level, geographical subpopulation, and virgin/mated status.

WEB OF INFORMATION

Female widow spiders are more likely to produce eggs fertilized by the first male with which she mates, creating strong selection for males to seek out and court virgin females. This is especially true because, although less frequently than their "widow" name implies, females sometimes attack and kill males before, during, or after copulation. If they survive the interaction, after mating, male *Latrodectus* sometimes destroy the female's web. This unusual behavior may impede—at least for a time—the female's ability to attract further males by limiting sex pheromone emission. This may also suit females, as mated females can manipulate the amino acid-derived sex pheromones on the silk to control mating attempts by males. Mated female silk contains a different chemical profile from unmated female

silk and elicits no courtship from males, thus communicating their lack of receptivity to males once their sperm stores are full, and saving the male the risk of potentially being eaten.

REDUCE DANGER AND MAXIMIZE PATERNITY

A female's recent feeding history is woven into her web, both in the web's silken architecture and in the chemical profile of the silk, and males pay attention. Mating with a well-fed female in good body condition is beneficial because she produces more eggs than a hungry female, so this information is useful for a courting male. Furthermore, as males may become their mate's next meal even before copulation, it can pay to know how hungry the female is: A well-fed female is less likely to attack a potential mate and engage in sexual cannibalism. Sure enough, males are more likely to court on the webs of well-fed females than starved females; after all, males that are eaten prior to mating don't pass on their genes and are consequently selected against.

→ Every country in which a species of *Latrodectus* resides seems to have a common name for the local species, presumably because it is venomous to humans. These names usually describe the spider's color or pattern for easy identification—for example, brown widow, black widow, white widow, redback.

CUPIENNIUS SALEI

Tiger bromeliad spider

Wind-catcher

SCIENTIFIC NAME	*Cupiennius salei*
FAMILY	Trechaleidae
BODY LENGTH	Females 1 ⅓ in (35 mm), males 1 in (25 mm)
NOTABLE ANATOMY	Large wandering spider with very long legs (reaching 4 in/10 cm)
MEMORABLE FEATURE	Jumping from leaves to catch flying prey at night

A sit-and-wait nocturnal hunter, the tiger bromeliad spider uses hairs to discern the slightest movement of wind that gives away an insect on the wing. Then, solely on the basis of the detected air disturbance caused by the unsuspecting flying prey, with pinpoint precision *Cupiennius salei* leaps up and snatches its meal on the fly in utter darkness.

A HAIRY BODY

Hiding during the day, *C. salei* typically lives on broad, unbranched leaves, such as bromeliads or banana plants. The leaves make for a springy platform from which to leap into the air to catch flying prey. Both substrate-borne vibrations and disturbances in the surrounding air are detected by inordinately sensitive mechanoreceptory hairs, which trigger the correct behavioral response to the detected situation, such as confronting a rival or escaping a predator.

Cupiennius have about 1,000 stiff hairs on their legs that are used to detect airflow. Called trichobothria, these hairs are arranged in groups of differing lengths. Each hair's length determines its best response to a different pressure load, deflecting in turn with stronger forces, or responding with mechanical sensitivity to miniscule levels. These hairs are thought to be the most acute sensory system known—it has been calculated that they respond to one hundredth of the energy contained in a single photon of green light! By having trichobothria of different lengths sensitive to and deflecting at different flows or air pressure, these spiders can calculate

the distance, direction, and speed of an insect flying nearby, and use this information to calculate where and when to jump into the night air to catch their prey.

A MECHANOSENSORY MINEFIELD

In addition to the trichobothria, *C. salei* uses hundreds of thousands of tactile hairs, stimulated by direct contact, for mechanoreception. Tactile hairs are not designed to deflect. Instead, they bend, making them both strong and sensitive to the tiniest flection. As if this mechanosensory world were not enough, about 3,500 "slit sensilla," or slits in the exoskeleton located at the joints, form sensory organs informing the spider of external vibrations of the leaf, such as the approach of a potential prey or predator, but also the pressure the spider itself is exerting on its cuticle.

Mid-air ambush
Frames of a video showing how *Cupiennius salei* reacts to the airflow produced by a tethered fly, from undetected at a distance of over 12 in (30 cm), to initiating a jump and leaping to catch the prey in mid-air once the insect is closer.

→   The tiger bromeliad spider's sensory systems have been described as "masterpieces of engineering" resulting from 400 million years of evolution.

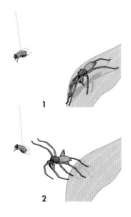

Paradise spiders

Serenading sweetheart

SCIENTIFIC NAME	*Habronattus dossenus*
FAMILY	Salticidae
BODY LENGTH	⅕–⅓ in (5–8 mm)
NOTABLE ANATOMY	Striking coloration of face and legs of males
MEMORABLE FEATURE	"Singing" courtship behavior of males

Iridescent greens and oranges ornamenting male *Habronattus dossenus* are put to great effect during courtship displays. One would be forgiven for thinking that the dynamic color vision-based components of courtship would be sufficient for males to convey their prowess to the rather drab females. However, sensory channels, such as pheromones and near-field sound, are likely extra modes of communication, and the use of seismic signaling comprises another staggeringly complex sensory modality.

SEISMIC SIGNALING

Living on exposed ground, the diverse jumping spider genus *Habronattus* is known for its colorful visual courtship displays and the complexity of its seismic "songs." Although we cannot detect it, mechanoreception through substrate-borne vibrations is a key sensory channel for communicating sex and species, and potentially more detailed information such as body condition. In *H. dossenus,* the seismic component of the male display is crucial for his mating success. Studying seismic songs and their relation to the concurrent visual displays of courting males requires coupling information gleaned from high-speed video and laser vibrometry. A laser is beamed at the surface on which the male is displaying, and the reflected (doppler-shifted) "echo" of the displacement of the surface caused by the signals is measured. Based on these studies, we now have a mechanism to understand *Habronattus* signaling patterns—one based on musical annotation.

ANNOTATING VIBRATIONS

Songs can consist of dozens of distinct vibrational signals, or elements, which can be repeated, even as new elements are added to the song as it progresses. A novel way to understand these songs is to annotate the vibratory oscillations as a musical score, allowing for distinct repeated motifs, each consisting of a variable number of elements to be discerned within each movement. An unresolved question is why this complexity in male signaling is necessary, since visual signals alone can provide species identification and other important information. Furthermore, there are major physiological limitations on information processing and memory experienced by females simultaneously assessing so much complexity in multiple sensory modalities. We have immense brains compared with the poppy-seed-sized brains of jumping spiders, and we have a name for this occurence: sensory overload. Theory suggests that signals that are reliable and simple should be favored, yet many female *Habronattus* seem to favor complexity, which may in turn be leading to further speciation.

→ Male paradise spiders often add a visual component to their displays to females by showing off their colorful knees, or a section of their third pair of legs (here in orange), during courtship dances.

PORTIA FIMBRIATA

Fringed jumping spider

Jack of all trades

SCIENTIFIC NAME	*Portia fimbriata*
FAMILY	Salticidae
BODY LENGTH	⅓–⅖ in (7–10 mm)
NOTABLE ANATOMY	Very large forward-facing eyes and tufty appearance of legs
MEMORABLE FEATURE	Hunting other jumping spiders

Portia fimbriata is fairly large by jumping spider standards. It uses extraordinary visual prowess to locate and hunt its dangerous, yet preferred prey: other jumping spider species. Having visually identified its prey, *Portia* stalks its target using a slow, robotic, "dead leaf drifting" pattern of approach, known as cryptic stalking. This may hinder the target spider's ability to detect *Portia* and either escape or attack.

MODULAR VISION

With supreme miniature telescopes for eyes, *Portia* also has superior visual skills to other jumping spiders, giving it a slight advantage over its prey. Jumping spiders have one large forward-facing pair of principal eyes and three smaller pairs of secondary eyes that surround the head. Combined, the eyes can see almost 360 degrees around the spider.

This wide field of view enables the secondary eyes to detect and track a source of motion, such as a predator or prey, with great precision. They can also see detail and so play a role in quickly categorizing moving objects and initiating responses, such as escape or prey capture. A target detected by the secondary eyes evokes a rapid turning response to bring the object of interest into the field of view of the principal eyes, which then process further detail.

A VISUAL MASTERCLASS

Jumping spider principal eyes have a unique anatomy. They consist of a lens, a long eye tube (see page 50), and a boomerang-shaped retina with a diverging pit located deep inside the head at the end of the eye tube. This increases the focal length of the system and forms a Galilean telescope, which magnifies the image but restricts the visual field to about 5 degrees. To compensate, the spiders can move the eye tube side to side and up and down by 30 degrees, allowing them to perform detailed visual scanning of objects over a much larger field of view and to discern intricate detail. Capable of discriminating individual items placed ¹⁄₂₅₀ in (0.1 mm) apart at a distance of 4 in (20 cm), *P. fimbriata* has visual spatial acuity unrivaled by any other studied animal of remotely similar size.

Fields of view

Fields of view of the eyes of most jumping spiders. Yellow: The large forward-facing principal eyes. Each yellow band represents the tiny field of view of each eye. Being movable about 30 degrees to either side, this small field of view is extended. Red: The overlapping fields of view of one pair of secondary eyes. Green: Another pair of secondary eyes completes the picture for a 360-degree world view.

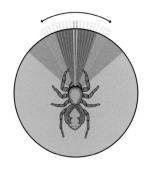

→ With amazing cognitive abilities, this spider is often regarded as the Einstein of spiders, but its problem-solving is somewhat slow.

DEINOPIS SPINOSA

Net-casting or ogre-faced spider

Night watcher

SCIENTIFIC NAME	*Deinopsis spinosa*
FAMILY	Deinopidae
BODY LENGTH	Females ½–⅔ in (12–17 mm), males ⅖–½ in (10–14 mm)
NOTABLE ANATOMY	Largest eyes of any spider (posterior median eyes ¼₀₀ in/0.07 mm)
MEMORABLE FEATURE	Use of expanding silk net to catch prey at night

As it hangs upside down with an expandable silk web spun between its front legs in the dead of night, the net-casting spider uses vision to ensnare prey walking beneath it by stretching the silken trap as it lunges, as if cast in a sci-fi movie. Web-building spiders generally have poor eyesight, but *Deinopis* is an exception.

Good eyesight can be as much a matter of spatial acuity as of sensitivity. Sensitivity is the amount of light an eye is able to capture and process, enabling vision in low light. Net-casting spiders are to nocturnal vision what *Portia fimbriata* is to diurnal vision, but unlike the enlarged principal eyes of the diurnal jumping spider, a massively enlarged pair of secondary eyes is responsible for *Deinopis'* visual accomplishments. While the nocturnal *Deinopis* uses sound to capture flying prey, vision is required to capture ground-based prey. Their huge (about ¹⁄₂₀ in/1.4 mm) posterior median pair of secondary eyes, earning them the name ogre-faced spiders, are thought to be 2,000 times more sensitive to dim light than our eyes. But how, when their eyes are so much smaller than ours?

THE DETAIL VS. NIGHT VISION TRADE-OFF

Each photoreceptor samples a specific part of a given image, so, if photoreceptors are small and densely packed, the image is sampled in detail, providing high spatial acuity. This is the case in the foveal region of our eyes (cone-based vision) and enables us to differentiate between strands of hair, but it comes at the cost of sensitivity. In low light, the limited number of photons reaching individual small photoreceptors makes them perform poorly, whereas large photoreceptors can capture enough photons to form a crude image or detect motion. However, this has negative impacts on visual acuity, as is the case with the poor resolving detail of our rod-based night vision. This leads to a classic trade-off: Spatial acuity improves as the ratio of photoreceptor diameter to focal length decreases, and sensitivity improves as the same ratio increases. The only potential solution is to have a very large eye, like our own, to accommodate both of those roles, but that is impossible in spiders, which are constrained by their exoskeleton to have small eyes. By having a short focal distance and large photoreceptors in their enlarged, yet relatively tiny eyes, *Deinopis'* posterior median eyes are well adapted to detect motion at night.

→ Being strictly nocturnal, during the day *Deinopsis spinosa* hides in plain sight by resembling a twig.

Algerian jumping spider

Stone dweller

SCIENTIFIC NAME	*Cyrba algerina*
FAMILY	Salticidae
BODY LENGTH	Females ⅛–⅓ in (4–7 mm), males ⅛–⅕ in (3–5 mm)
NOTABLE ANATOMY	Bright-orange head of males
MEMORABLE FEATURE	Hunting spiders in poorly lit areas

Unusually, the jumping spider *Cyrba algerina* visually distinguishes conspecifics and assesses its prey in both very dim and very bright light, despite not having atypically large eyes. Living under dim light conditions beneath stones in arid habitats, the Algerian jumping spider hunts web spiders in preference to insects as prey, and visually discriminates between prey under light conditions equivalent to early or late dusk. Using intricate predatory behaviors, *C. algerina* often hunts under stones, but using a visual system able to achieve sufficient spatial acuity for object identification under a range of lighting conditions, it can also hunt in more open and brightly lit areas above stones.

A COMPLEX PREDATOR

Possibly due to the dangerous nature of its preferred prey, *Cyrba*'s venom is specialized to rapidly subdue spiders compared with insects (12 times faster, in fact), and its predatory behavior is complex. Spider-specific predatory behavior includes using trial and error to derive vibratory signals which are plucked on the prey's web to gently coax the resident spider toward the awaiting hunter without evoking a predatory response from the web spider.

When it invades a prey spider's web, *C. algerina* also capitalizes on wind disturbances on the web, masking its own vibrations by using the wind as a smokescreen and rapidly advancing toward the resident web spider prey, while ignoring any insects caught in the web. These intricate

behaviors are visually mediated, irrespective of whether they are performed in bright sunlight or under ambient light levels associated with nocturnal species.

A BLENDED LIFESTYLE

Since the amount of light captured is usually related to eye size, species with small eyes face major difficulties. Compared to most jumping spiders, which live in bright light, the principal eyes of *Cyrba* have a short focal length, and its small photoreceptors have evolved strategies to merge photon capture, essentially making the photoreceptors larger and favoring sensitivity without fully compromising acuity. These adaptations may be beneficial for a jumping spider with a "blended" lifestyle: generally living and hunting under stones in the dark, but sometimes venturing above them in dramatically different light conditions. *C. algerina* illustrates how sensitivity seems to have been favored over spatial acuity, allowing this species to minimize the constraints imposed by its particular microhabitat.

→ Populations in different regions specialize at eating different types of spiders, and each population learns the specific odors of the locally abundant prey to facilitate prey detection.

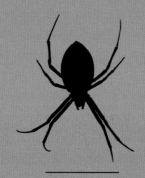

SILK &
WEB-BUILDING

An unusually versatile material

Sensing an oncoming storm due to the electrical fields in the atmosphere, a baby spider (spiderling) climbs to the tallest area of foliage it can reach and deploys—perhaps for the first time in its life—silk. The silk filaments waft into the air, caught by the breeze and the electrical fields in the atmosphere, and the spider is lifted away to new areas, sometimes hundreds of miles away, to make a new home.

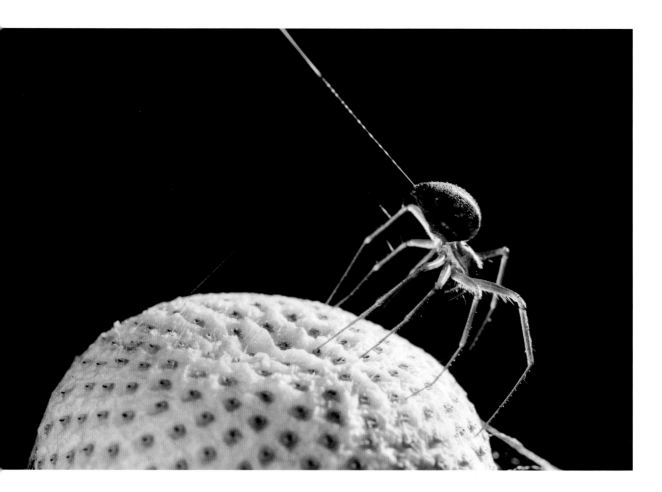

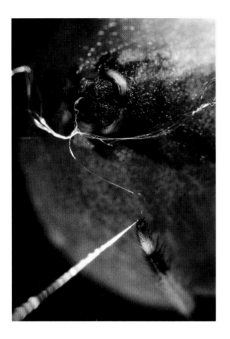

This tiny spider, barely visible to the human eye, has detected the electrically supercharged atmosphere due to the storm through its trichobothria, now acting as electromechanical receptors. This has triggered a behavior known as ballooning, whereby the spider seeks an exposed area and, standing on tiptoe, lifts its abdomen in the air and releases silk to prepare for its first flight.

All spiders produce silk, which, being both tough and flexible, is very versatile. With a few adjustments, silk can be made to have slightly different properties for different tasks. Consequently, a given spider may have multiple types of silk glands in their abdomen, each producing a different type of silk. Within the gland, the silk is a protein-rich liquid that flows through a duct that removes water from the liquid. The spider uses its hind legs to pull silk out from its highly mobile spinnerets, creating tensile stress. This tensile stress further solidifies the viscous fluid by causing the protein molecules in the silk to align and form strong bonds with each other, transforming it into a solid.

MULTIFARIOUS SILK

Spiders have capitalized on the versatility of silk, using it to disperse or build webs, protect their eggs, or line their burrows. They use silk to communicate and to subdue prey. They use silk as a sensory system. Males produce sperm in a different organ to the one used to transfer sperm, so they build "sperm webs" in which they deposit the semen to fill the palps with sperm for future use. In short, silk plays a fundamental role in all aspects of a spider's life, and this requires them to produce silks with distinct physical properties to perform different functions.

↑   Spiders can have two, four, six, or eight spinnerets, each ending in tiny spigots to provide the tension required to turn the internal liquid silk into the extruded solid threads.

←   Ballooning is a key dispersal mechanism for spiders, allowing them to colonize new lands. It is believed much of New Zealand's spider fauna arrived this way from Australia, thousands of miles away.

Even when deployed in the "traditional sense" of building a web, spiders produce an array of silks corresponding to the different roles they play within the web, such as elastic silk that is stronger than steel to anchor the web to the surroundings, or sticky silk to ensnare prey. An orb web contains up to five types of silk. Tough dragline silk, produced by the major ampullate gland, creates the frame in which the hub will be built, as well as the radial threads that crisscross through the final structure and provide structural support, like spokes on a bicycle wheel. The spirals within the web are produced by silk from the flagelliform gland; this is coated in silk with glue droplets produced in the aggregate gland to make the "capture zone" of the web sticky.

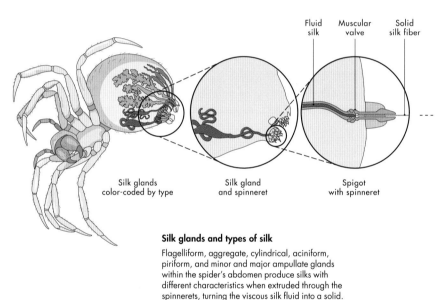

Fluid silk   Muscular valve   Solid silk fiber

Silk glands color-coded by type

Silk gland and spinneret

Spigot with spinneret

Silk glands and types of silk

Flagelliform, aggregate, cylindrical, aciniform, piriform, and minor and major ampullate glands within the spider's abdomen produce silks with different characteristics when extruded through the spinnerets, turning the viscous silk fluid into a solid.

Minor ampullate silk from its corresponding gland can be used as reinforcement silk for the spirals, or as scaffolding during construction, which is removed on completion. Finally, the silk cementing the web to its anchor points is made using pyriform silk spun as an attachment disk. Say that the spider then catches prey: She wraps it using aciniform silk. She also uses aciniform silk to make a soft lining for the protective case in which she houses her fertilized eggs until they hatch, coating the case with tough cylindriform or tubiliform silk for protection.

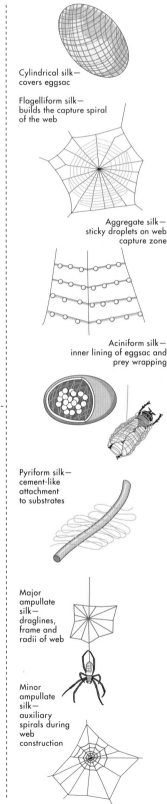

Cylindrical silk—covers eggsac

Flagelliform silk—builds the capture spiral of the web

Aggregate silk—sticky droplets on web capture zone

Aciniform silk—inner lining of eggsac and prey wrapping

Pyriform silk—cement-like attachment to substrates

Major ampullate silk—draglines, frame and radii of web

Minor ampullate silk—auxiliary spirals during web construction

→ Sperm is produced within the abdomen and is then ejected from a genital opening, or gonopore, onto a silk mat (sperm web) for transfer to the palp, or copulatory organ.

FINE FIBERS

Araneomorph spiders can also be divided into cribellate (wooly) or ecribellate (sticky) spiders. In addition to having spinnerets, cribellate spiders extrude hundreds of very fine, dry-silk fibers—and some thicker ones for good measure—through a sievelike plate (cribellum). These are combed out of the cribellum with stiff bristles on the rearmost pair of legs, creating the characteristic fluffy woolen appearance of cribellate silk. While ecribellate spiders use aggregate sticky silk to catch prey, the prey of cribellate spiders largely get caught on cribellate silk through entanglement, as if on Velcro.

Spider spinnerets

Spiders can have several pairs of spinnerets at the rear of the abdomen. Each spinneret consists of many individual silk-producing spigots connected to individual silk glands.

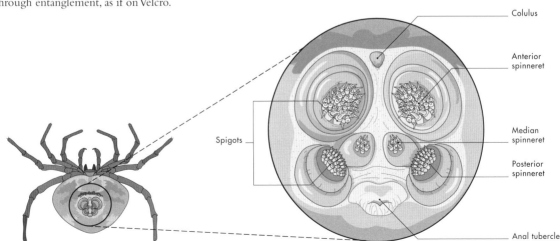

Colulus

Anterior spinneret

Median spinneret

Posterior spinneret

Spigots

Anal tubercle

From the ground to the air

An almost perfectly rounded orb web sagging in the dew and reflecting the early morning sun is an evocative image, but orbs are just one of many shapes that are used to create webs. Aerial webs have evolved from the ancestral use of silk to line burrows, as still used by tarantulas, other mygalomorphs, and liphistiid spiders. Radiating silk lines extending from a burrow can be imagined as a ground-based web.

As insects evolved the capacity to fly, webs rose to the occasion: The evolution of aerial webs—horizontal, then vertical, coupled with sticky silk—corresponds with the evolution of insect flight, and the biggest spurts in spider diversification follow similar diversification in insects. More recent spider lineages are increasingly losing web-building capacity: They limit their use of silk to other functions, such as egg protection and communication.

Spider web and silk diversity

Spiders make the most of the habitat in their use of silk: from ground-based spiders using silk "funnels," to those that use vegetation or human-made structures to support angular, horizontal, vertical, expandable, collapsible, or spring-loaded varieties of silken prey-catching devices.

A: Horizontal sheet web, linyphiid spiders, B: Detached cribellate web, e.g., *Deinopis*, C: Reduced orb web, e.g., *Mastophora*, D: Reduced cribellate orb web, e.g., uloborids, *Philoponella*, E: Orb web, e.g., *Nephila, Argiope*, F: Social spider orb web with retreat in contact with support structure, e.g., *Stegodyphus*, G: Tent web, e.g., *Cyrtophora*, H: Ladder web, e.g., *Telaprocera*, I: Ground-based "web" of burrowing spiders with extruding silk triplines, e.g., *Atrax*, J: Gumfooted web, e.g., *Latrodectus*

↗ Orb webs can be vertical, horizontal, or at an angle. This does not depend on the habitat structure around the web being used for the anchor points, but on spider species.

→ A jumping spider uses silk to build a tentlike retreat, or cocoon (often in rolled-up vegetation), in which to lay eggs, escape bad weather, rest, and molt.

GRAND DESIGNS

Web architectures mirror the diversity of functions fulfilled by spiders in their ecosystems, so come in many shapes and forms. Webs can also differ between individuals of the same species and webs made by the same individual. However, there are broad forms of web architecture. Burrowing spiders might have more than radiating lines; they may also extend a trampoline-like sheet over the ground surrounding the burrow entrance, from which they don't stray far. Similarly, derived forms of orb webs often reflect a change from generalist traps to catch a variety of prey, to smaller, more specialized traps for specific prey. In some groups, only a slice of an orb is built or only the bottom half; in other cases, it has been reduced to a web held by the legs and either expanded, as in ogre-faced spiders (see page 62), or used as a lasso, such as in bolas spiders (see page 92). Cases where the orb has been expanded also exist: Ladder web spiders have extended the upper and lower portions of the vertical orb web, while in horizontal webs like those of *Cyrtophora*, additions above the web— possibly for structural support, protection, or to knock intercepted prey to the bottom sheet below—make the now three-dimensional (3D) structure appear like a tent.

3D ROBUSTNESS

Three-dimensional web types are more common than two-dimensional (2D) webs; they are also typically more permanent and require silk that can withstand the elements for long periods. These webs can be broadly classified as: those that have a horizontal hammock-like sheet, typical of linyphiids (dwarf, sheet web, or money spiders); webs without a clear horizontal sheet but with an array of sticky strands, below which they are attached under tension to the ground (gumfooted webs); and irregular webs without sticky trapping lines (tangle webs). Gumfooted webs, like those found in widow spiders, trap prey walking along the ground. As the prey walks into one of the lines extending to the ground, it breaks the thread—on which it is now stuck—and is lifted into the air to be reeled up by the waiting spider.

WEB TYPES

Dismiss the notion that all spider webs
are two-dimensional, flat, and round—the
so-called "orb web." The diversity of webs
mirrors that of their makers. Even among
orb-weaving spiders, some species will
simply "delete" a section of the orb. These
are aptly called missing-sector orb-weaving
spiders. Some create webs that are held in
their legs, and some of these are expandable
—a complex silken rubber band snaring net.
Other species make webs that look like the
main bigtop tent in a circus in miniature.
These three-dimensional tent webs have
different dynamics to two-dimensional, or
planar, webs, and so the mechanisms by
which they ensnare prey also differ from
planar webs. Some webs contain sticky silk
that houses gluelike droplets to prevent the
escape of trapped prey, while other silks
act like Velcro. Webs can be the size of
a thumbnail, or larger than a truck. Webs
are spiders' architecture, and, as with our
human architecture, there are many possible
solutions to a problem—in this case, how to
catch food.

↗   No one really knows why so many
spider species expend extra energy to
"decorate" their webs with myriad patterns,
which can be species-specific, age-specific,
or highly variable.

→   Trees can become covered in silken
webs from spiders—even nonsocial
species—escaping temporary floods.

→→   Web spiders also prepare dense
silken nests to protect their eggs. These are
often hidden in a corner of the web, or in
a leaf within the web.

Flexibility in silk

Spider silk is different to that produced by other organisms (such as silkworms, which confusingly are the larvae or caterpillars of moth species that produce silk used in textiles). Spiders have many more types of silk, and the threads are much finer. In fact, the diameter of some types are more than 20 times narrower than a human hair. The only reason we can see spider silk at all (unless it is dew-laden) is because of the way it reflects light.

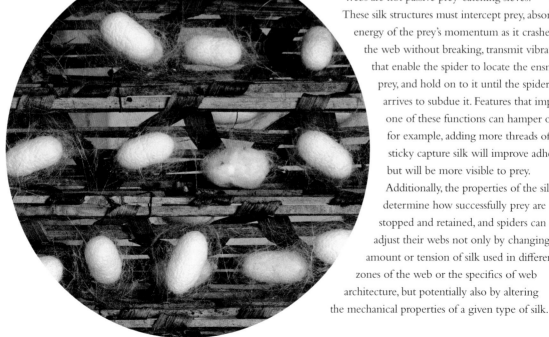

So, spider silk is really *really* thin, and yet webs can catch very large prey, including birds. This begs the question of how spider silk stays intact on being hit by anything at all, and manages to be both firm and stretchy. Steel is firm and will stop a bird in its tracks, but it has little elasticity. A rubber band is elastic, but when stretched too much it breaks easily. Silk is unmatched when it comes to how effectively it combines these two opposing properties in a single material.

Webs are not passive prey-catching sieves. These silk structures must intercept prey, absorb the energy of the prey's momentum as it crashes into the web without breaking, transmit vibrations that enable the spider to locate the ensnared prey, and hold on to it until the spider arrives to subdue it. Features that improve one of these functions can hamper others; for example, adding more threads of sticky capture silk will improve adhesion, but will be more visible to prey. Additionally, the properties of the silk determine how successfully prey are stopped and retained, and spiders can adjust their webs not only by changing the amount or tension of silk used in different zones of the web or the specifics of web architecture, but potentially also by altering the mechanical properties of a given type of silk.

↑   Silk is so strong that despite often being invisible to the naked eye, it can stop and restrain large animals, such as this common noddy tern in the Seychelles, without breaking.

←   The ancient art of silk farming: Silkworms can be raised on wooden frames to reel the silk filaments from their cocoons.

Consider the difference required of the material if its function is to absorb energy or prey momentum, as in an orb web compared with gumfooted webs, which on contact with the prey, snap and catapult it up toward the spider. The pyriform silk disk used for the attachment to the substrate is tough and durable for the orb web, while the gumfooted web has weak attachment points to the "gumfoot" threads, making these spring-loaded to be easily triggered by a walking insect.

SILK VARIABILITY

Like all materials, the mechanics of silk can be described in terms of: how much energy is required to stretch a thread to breaking point (toughness); how much force the thread can withstand given its diameter and other characteristics of elasticity; or the increase in length at breaking compared with its original length. Spider silk toughness is unrivaled, but it is also variable, not only depending on the type of silk produced, but also within a given type of silk.

Evidence suggests that spiders can modify these parameters depending on circumstances such as humidity, habitat structure, or wind conditions, or depending on their own diet, age, size, and the types of prey available.

Web-building also changes with age and experience. In some species, as they get older, spiders change the shape of the web in stereotypical fashion, but in others, spiders use previous experience of where a web has been especially successful at intercepting prey

↑ Spiders use leg and body movements to position themselves for correct silk placement and tension during each of several building stages, making the web a direct record of their countless movements as they build.

↗ Carefully clipping on to different threads radiating from the hub, the spider is alerted to the intensity and directionality of a given disturbance by the smallest vibration.

  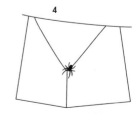

Web construction

Web construction goes through several phases. 1. An initial bridge thread is attached between two anchor or attachment points. 2. The beginnings of a radial thread. 3. Radial threads are tensioned to a new anchor point. 4. Further anchor points are made to complete the frame. 5. Further radial threads are attached to the frame. 6. The beginnings of the spiral threads. 7. Spiral threads are added. 8. The densely threaded central capture zone is often coated with adhesive silk.

to modify the web to maximize catch rate in that area, or to reduce wind drag in areas especially susceptible to wind damage. Sometimes there are seasonal changes in the types (and sizes) of prey available, and web-building reflects those changes: for example, reducing costs by using more widely spaced threads at times when bigger prey are especially common. These costs may well regulate much of web-building behavior and are not trivial. To produce enough silk a spider must be well-fed, and the silk used, in turn, will factor into how well-fed the spider is. There is also the cost of the time taken to create the web: Webs may take many hours to create, yet some species will take them down daily, recycling the protein-rich material by ingesting it during deconstruction before beginning the laborious process all over again.

GENIUS GENES

The "blueprint" of web architecture and silk type used by a given species of spider is genetically encoded. Although complex, the web-building behavior of orb-weavers is quite stereotypical, in that from birth, spiders employ subroutines in web construction that typically vary only to accommodate the parameters imposed on a web by the habitat, such as available space. Essentially, orb web construction, which is the best studied, consists of a sequence of fairly well-defined stages of assembly, starting with the frame, followed by the radii, then the spirals and their temporary reinforcement scaffolding, and ending with the dense area of the sticky capture spirals near the hub of the web. However, spiders can alter these subroutines, for example, by adding more radii to the completed web.

This suggests that while the blueprint for a web is present, spiders have the flexibility to improve on their design, presumably based on need. Interestingly, because it progresses in stages, web design may involve memory of where the spider is in the sequence of thousands of paces and connecting strands required for its construction. The spider may not literally remember within its brain where it is in the process: The web itself is a physical memory of what has been achieved—touching its strands may be all that is required.

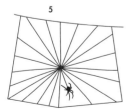
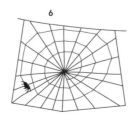
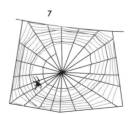
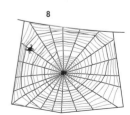

Spreading the word through silk

If only webs were so simple as to be merely external memory devices used for catching prey. They are, of course, external sensory systems, too. As such, they play a crucial role not just in detecting caught prey, but in communication between spiders connected by a thread.

In reality, it is not as simple as a one-to-one connection with a single line. Web architecture precludes this. In an orb web, vibrations from a source will propagate and dissipate (or lose energy) accordingly through any number of interconnected threads, and the energy will be further affected by how many nodes, or points of connection between strands of silk, it intersects. This is in a regular, nicely geometrical web, but consider how a spider might even begin to localize the source of a vibration detected among the mess of silken threads that form tangles or cobwebs that have no apparent geometrical regularity at all. This would be like trying to detect exactly which line was pulled where in a tangled mess of parachute lines. By modifying web geometry or architecture, as well as the properties of the silk and the tension under which the silk is strung, spiders, of course, can figure that out, and plenty more besides.

← With legs attached to different strands of silk, the spider can read vibrations from every direction. Based on the strength, frequency, and type of vibration, it decides what to do.

ON THE SAME WAVELENGTH

Without getting into specifics, vibrations propagate as waves through the medium. In silk, they propagate as waves with different displacement angles, or axes. Some waves have their axis of displacement along the thread (longitudinal waves), whereas others (transverse waves) are perpendicular to the axis of the thread. Each wave type therefore differs in the speed at which it propagates through the web, how it spreads, and how it dissipates. Waves are also characterized by their given frequency, which is the number of waves, or wavelengths, per second, measured in hertz (Hz). A courting male might pluck a web, producing waves at a given frequency of, say, 120 Hz to signal to the resident female, while a wasp buzzing nearby might produce wavelengths of 200 Hz, providing the resident spider with a cue that danger might be flying nearby.

Web geometry influences the propagation of all these types of waves as they spread like irregular ripples across the structure. Orb webs seem to be especially good at transmitting longitudinal waves, while tangle webs appear to be pretty good at transmitting both longitudinal and transverse waves. By altering the geometry of their specific type of web, spiders can optimize the transmissibility of both specific wave types and the specific frequencies that best match the vibration frequencies produced by items of relevance to the spider, such as prey, predators, or potential mates.

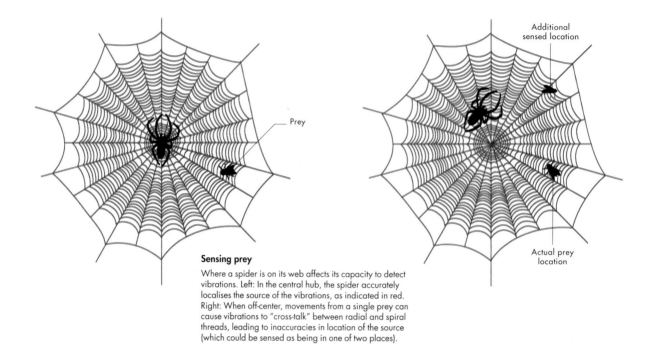

Sensing prey

Where a spider is on its web affects its capacity to detect vibrations. Left: In the central hub, the spider accurately localises the source of the vibrations, as indicated in red. Right: When off-center, movements from a single prey can cause vibrations to "cross-talk" between radial and spiral threads, leading to inaccuracies in location of the source (which could be sensed as being in one of two places).

Additionally, during the extrusion of silk, spiders can change the speed at which they reel out the silk. This influences silk stiffness because faster speeds lead to increased order within the silk protein structure, or stronger bonds between the molecules, in turn leading to higher stiffness. Very rigid webs may be great in windy conditions, but may be quite poor at propagating vibrations, and spiders will adjust their silk accordingly. A spider can also manually adjust the tension of each strand of silk. It can cut lines and redo them under higher tension, or it can pull strands and "retie" them at the nearest node, or it can simply "hold" them for a little while if it detects something that might be of interest.

VIBRATION DECODING

Using its slit sensilla and other mechanoreceptors, orb web spiders are especially sensitive to longitudinal vibrational waves, especially if they stand on their webs such that their tiptoes (the leg tarsus) are almost perpendicular to the fibers on which they rest. This means that they can also adjust their posture to maximize their sensitivity if something potentially interesting is detected. By comparing the strength of the vibrations, or amplitude, detected between the different sprawled-out tarsi, the spider can potentially determine the location of the vibrations, since stronger vibrations should signal the direction of the source— although this is disputed because the extreme speed with which vibrations propagate through webs may be too fast for neurons to process the theoretically detectable differences between the tarsi. Yet, spider sensory systems frequently defy expectations based on theory.

↑ Courting a female much larger than himself, a male (top) must signal very carefully on her web, ensuring she responds in the best possible way. Appropriate communication here is literally a life or death situation for the male!

← Spider legs end in the tarsus, from which emanate one or more movable hooklike "tarsal claws." The spider uses these to attach themselves to surfaces or delicately maneuver silk.

So what information can a resident spider obtain from, say, a courting male using vibratory signals on her web? Although many different behaviors associated with producing vibratory signals have been described for many spiders—such as the plucking of threads with the legs, shuddering the abdomen against the web, or full-on body bounces—little is known about what information is gleaned by the resident. Some signals have clearly been correlated with body size, suggesting that females would certainly be able to determine male size, which is a potential indicator of his condition. Additionally, it is highly likely that she will be able to discern if the signaler is a courting male of her species, due to the specific frequencies produced by the signals. Undoubtedly, the female is gleaning more information than that, but we are limited by technology in discovering what just yet.

Red-legged golden orb-weaver

Female giant

SCIENTIFIC NAME	*Trichonephila inaurata* (formerly *Nephila inaurata*)
FAMILY	Araneidae
BODY LENGTH	Females 2 in (50 mm), males 1 in (25 mmm)
NOTABLE ANATOMY	Very large spider with red bands on its long legs (leg span up to 6 ½ in/160 mm in females)
MEMORABLE FEATURE	Building golden webs spanning several yards

The largest fabric item made from spider silk is a cape measuring 4 ft by 11 ft (1.22 m by 3.35 m). Handwoven from the silk of more than a million *Trichonephila inaurata* spiders from Madagascar, the cape and a matching shawl took years to make. About 80 people used a centuries-old technique of a hand-drawn reel to extract the precious silk from the spiders before releasing them back into the wild.

When it was finally complete, in 2009, the clothing was exhibited in museums around the world. Beautifully embroidered, light and resilient, its lustrous golden color is most striking: *Trichonephila* are called golden orb-weavers due to the characteristic yellow-tinged silk produced by the supersized females.

GIANT WEBS

Exhibiting the female gigantism typical of its genus, *Trichonephila inaurata* females have a leg span of 4 in (10 cm) and build orb webs with a diameter of up to about 5 ft (1.5 m). Several tiny males reside on the fringes of the web, waiting for an opportunity to mate with the resident female. Webs of this size require elevated anchor points. It is common to see these webs spun from telephone lines to the ground. Although they typically eat insects caught in the webs, reports of small birds getting ensnared are not uncommon, possibly because the webs are high enough to be in their flight path.

CRAZY SILK

Most research on *T. inaurata* is on the structure of the silk produced by the major ampullate gland, which, at about 8 μm diameter, is about 30 times narrower than human hair, but very thick by spider standards. Silk has phenomenal tensile strength and springiness. These attributes together yield toughness well above that of almost all known materials. For example, although there is considerable variability in silk characteristics between species—and even within an individual spider based on what the silk is used for—in *T. inaurata* the tensile strength of major ampullate gland silk is similar to that of high-tensile-strength steel, but its toughness is 20 to 40 times higher. These mechanical properties of spider silk make it of considerable interest to materials science and biotechnology. Adding to this, spider silk can supercontract, or return to its original state when immersed in water, irrespective of the loading it may have experienced: Soak and it's like new! This array of remarkable mechanical feats is due to the layering of proteins within the silk, but the nanoscale properties of silk remain largely unknown and probably account for much of its exceptional performance.

→   Multiple smaller males will court a large female. Emerging from the egg smaller than a pinhead in size, she grows up fast, but despite her large size, she only lives for about 12 months.

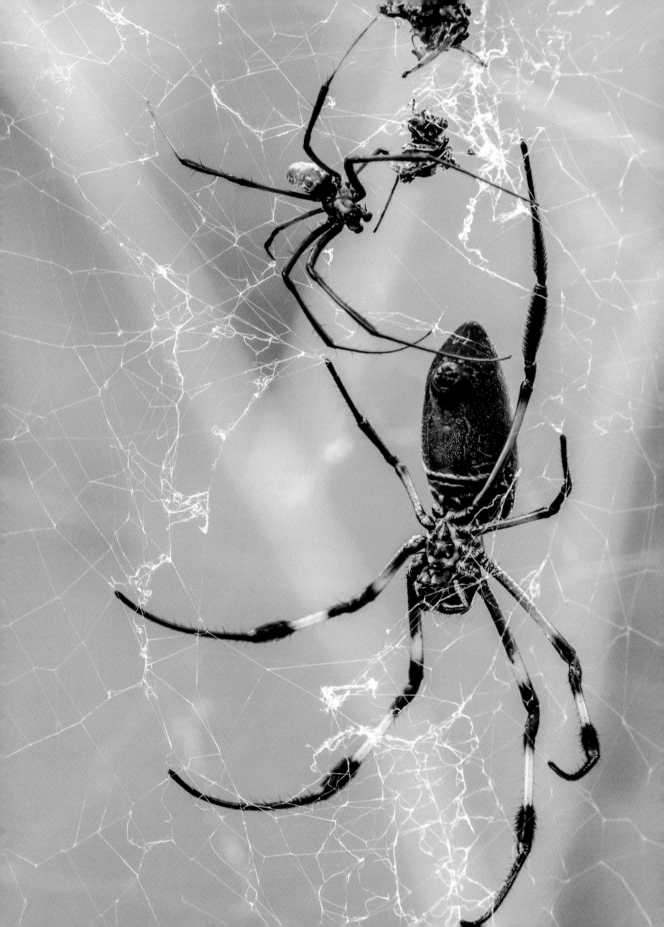

Tropical tent web spider

Nice neighbor

SCIENTIFIC NAME	*Cyrtophora citricola*
FAMILY	Araneidae
BODY LENGTH	Females ⅖ in (10 mm), males ⅛ in (3 mm)
NOTABLE ANATOMY	Cryptic, "lumpy" appearance
MEMORABLE FEATURE	Aggregates in colonies

We can all recognize an orb web, but about 90 percent of webs are 3D mesh-like structures. Scaled up to our size, these would be many stories high. Perhaps the most recognizable 3D webs are those of tent web spiders. They build a dense horizontal sheet within a silken mesh. These 3D webs are relatively permanent and must be resilient to damage.

Based on laser scans of the webs of *Cyrtophora citricola,* we are unraveling the interactions between the nonlinear mechanics of silk, fiber length, and connectivity, or the number of fibers to which each node is connected. This is being used to develop machine learning neural networks and signal transduction and resiliency in fiber networks, such as cell phone connectivity after a disaster that might affect cell towers.

COMPLEX LIVES

With the possibility of ameliorating costs of predation and of increasing food capture, sociality (sociability) has many benefits, but the increased competition for mates or food is a substantial cost to this approach. Although social behavior is uncommon in spiders, *C. citricola* typically live in aggregations of many—sometimes thousands—of individual webs. Although they do not cooperate to catch prey, aggregating seems to increase the overall efficiency of prey

capture. However, these spiders can also monitor the demographics of the colony—and if the colony outgrows itself, the spiders detect that future reproductive effort will decrease in the larger aggregation, and they will disperse to live alone or in smaller groups with less competition.

COMPLEX WEBS

3D tent webs rely on intercepted prey bouncing within the mesh to reach the tent sheet under which the spider sits. This is less efficient than 2D planar webs, but those require a lot more maintenance and are less effective barriers against predators. The beauty of 3D structures is in their resilience. Their ability to absorb energy is not solely due to the elasticity of the silk, but to the actual structure of the web. Consequently, the tensile strength and toughness increase as a function of the density of the web because the load is distributed across many silk fibers. This redundancy is crucial for its robustness.

→ These spiders can live as solitary individuals or can be found in colonies of hundreds of spiders, but if their "personal space" (the prey-capture part of their own web) is invaded, they become aggressive.

ANELOSIMUS EXIMIUS

South American social spider

Family lover

SCIENTIFIC NAME	*Anelosimus eximius*
FAMILY	Theridiidae
BODY LENGTH	⅛–¼ in (4–6 mm)
NOTABLE ANATOMY	Brightly colored red-orange body
MEMORABLE FEATURE	Extreme sociality

Less than 0.05 percent of spider species are permanently social, and the lowland rainforest-dwelling theridiid spider *Anelosimus eximius* is one of these. It seems that the ability of these ¼ in (6 mm) spiders to catch large prey drives their social lives, so once again, it all comes down to the web. Nesting together in colonies that can number tens of thousands of individuals, the spiders' web mass draped over trees can span several yards in any given direction and reach volumes of over 131 yd³ (100 m³).

SOCIAL SPIDERS AND SEX

Males comprise less than 1 percent of the individuals in a colony and roughly 70 percent of adult *A. eximius* females do not reproduce, helping instead with colony tasks. The extreme female-biased sex ratio is not due to male dispersal: Although their heterogametic reproductive system should yield similar numbers of male and female offspring, by mechanisms unknown, the embryos are already female-biased. Coupled with a relative lack of dispersal, there are high levels of inbreeding within each colony. Being related to colony members facilitates cooperative behavior such as web maintenance and brood care. This is because in an environment where few can reproduce, helping others increases the reproductive chances of an individual that shares many of the helper's genes (often more than full siblings would share), and so indirectly benefits the helper.

COOPERATIVE PREY CAPTURE AND OPTIMUM COLONY SIZE

Although about 90 percent of the prey ensnared in their webs is small, cooperation allows *A. eximius* to capture and simultaneously feed on prey much larger than themselves, contributing about 75 percent of the energy intake of the colony. Cooperative foraging also allows these spiders to ensnare and subdue increasingly large insects with increasing colony size—but there are scaling issues. Each colony member contributes a similar volume of prey interception area to the web, but it is the outer surface of the web that determines how much prey is caught.

For any given 3D shape, this surface area will decrease as a function of colony size, leading to competition for food as the colony increases. As a result, prey consumption per capita is optimal in mid-sized colonies of about 1,000 spiders. Intriguingly, populations at the elevational limits of their range are less social than those at the core, and this may be because larger insects are less frequently encountered at higher elevations. In this case, prey size may be the limiting factor on both sociality and the distribution of this species.

→ These spiders cooperate to build, defend, and clean their colony web, and to subdue their prey. While a single spider lives for less than a year, colonies live on for many years.

Maude's ladder web spider

Tree trunk guardian

SCIENTIFIC NAME	*Telaprocera maudae*
FAMILY	Araneidae
BODY LENGTH	Females ⅖ in (10 mm), males ¼ in (6 mm)
NOTABLE ANATOMY	None on body; elongated web
MEMORABLE FEATURE	Elongated web built along tree trunks

Often thought to be tediously stereotypical, web-building behavior is in fact often characterized by extreme plasticity. A ladder web is a type of web built by some orb-weaving spiders that is more than twice as long as it is wide, and it can be up to seven times as long. From a functional perspective, ladder webs might seem like a poor way to catch prey, because orb web spiders sit at the central hub of their typically rounded webs.

Prey intercepted farther away from that hub therefore have a greater chance of escape before the spider reaches them. It is a long way up or down from the hub in a ladder web, yet ladderlike webs have repeatedly and independently evolved among several orb web spiders.

CONVERGENT EVOLUTION

In some cases, the elongated shape of ladder webs is an adaptation to specifically catch moths. Being scaly, moths can often escape from webs, since they lose scales and consequently tend not to stick to webs. With ladder webs, scales flake off as the moths tumble down the elongated web until they finally get stuck, resulting in extreme specialization for moth capture in these webs. These moth-catching ladder webs, such as those of *Scoloderus* species, are aerial webs built in open spaces. In a case of similarity of form but not of function, convergent evolution has led to other ladder webs. The webs of Australian ladder web spiders *Telaprocera maudae* are built about ¾ in (2 cm) above tree trunks, meaning that

these spiders are limited in how wide their webs can be. Illustrating that the web is a flexible response to the available microhabitat, *Telaprocera* web architecture matches the tree trunk on which it is anchored: Wide trunks have shorter ladders than narrow trunks.

OPTIMIZING WEB ARCHITECTURE

Modeling shows that as horizontal space becomes limited, the optimal web shape is an elongated form. Although this results in the reduction of prey caught compared with webs built without physical constraints, it is an improvement on a typical radial web built in the same area. In other words, ladder webs make the best of a suboptimal situation. Intriguingly, there are parallels in spider behavior: Moth-catching aerial ladder web spiders build their webs every day during the day and are restricted to intercepting prey at night, which is suitable for catching moths. In contrast, *Telaprocera* webs are built at night and are only replaced every few days. With few moths, and a diversity of both day- and night-active prey types being ensnared, *Telaprocera* webs maximize foraging potential by intercepting prey at any time.

→ Living in closed-canopy rainforest, *Telaprocera maudae*'s flexible web-building behavior can be observed when it is given plenty of space: In these cases, webs are almost circular.

Bolas spider

Cowgirl arachnid

SCIENTIFIC NAME	*Mastophora hutchinsoni*
FAMILY	Araneidae
BODY LENGTH	Females ⅖ in (10 mm), males ¹⁄₁₂ in (2 mm)
NOTABLE ANATOMY	Looks like a bird dropping
MEMORABLE FEATURE	Bolas swinging predatory behavior

Bolas spiders like *Mastophora hutchinsoni* hunt using a single strand of silk with one or more large droplets of sticky silk (bolas) attached to it. Holding this lasso with a leg, the spider swings it to catch prey. This may seem futile, but the spiders produce a scent that mimics that of female moths, making the bolas an attraction magnet for male moths in search of females.

CHEMICAL WARFARE

A single horizontal line strung between twigs or leaves with a spider attached to it may seem like a web in bad shape. However, an adult female or older juvenile bolas spider is in its element here. Closer inspection reveals that it holds a short, vertical thread of silk with an alluring bolas. The production of the bolas can be stimulated by the vibrations of the wing beats of approaching moths. The viscous glue of the bolas seems especially well adapted to capture moths, which due to their scaly wings rarely get stuck on traditional webs. The odor emitted by the spider mimics the species-specific sex

pheromones of female moths, attracting male moths toward the source of the smell. Once within striking range, the spider swings the bolas to snare the moth before drawing it in or climbing down the thread to paralyze it.

SYNTHESIZED COCKTAIL

Each *Mastophora* species appears to specialize in preying on a limited number of species of moths. For example, *M. hutchinsoni* preys on four species. The constituents of the sex pheromones are known for three of these moths and have no chemical similarities. This implies that bolas spiders are flexible in their ability to biosynthesize a diverse set of perfumes mimicking the pheromones of their prey. Moths have distinct diel periods of sex pheromone emission and male responsiveness that last only a few hours. Perhaps the spiders can change their chemical lures during the night to target species active during those periods. Not only are bolas spiders flexible in the production of chemicals to attract prey, but they exhibit predatory versatility depending on age and sex.

Bolas spider prey capture

The ultimate lasso artists: 1. The spider hands off a bridging thread and dangles a strong thread of silk, ending in a sticky droplet, between her legs. 2. On detecting a nearby moth, she swings the thread and a moth becomes glued to the droplet. 3. When the prey is reeled in, the spider proceeds to wrap it for future consumption.

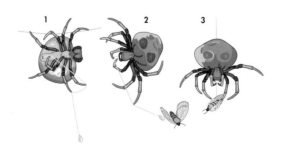

→ The larger female bolas spiders (seen here) hunt with a bolas. Males and juveniles do not produce the bolas, but *Mastophora phrynosoma* males do attract psychodid flies (presumably also through chemical mimicry), which they catch with their legs.

Leucauge argyra
Brainy builder

SCIENTIFIC NAME	*Leucauge argyra*
FAMILY	Tetragnathidae
BODY LENGTH	Females c. ⅛–⅖ in (4–10 mm), males c. ⅛–¼ in (4–6 mm)
NOTABLE ANATOMY	Yellow and silver markings on abdomen
MEMORABLE FEATURE	Makes a horizontal web in slanted position and rests in the middle of the web with its underside facing upward

There is a limit to how far neurons can miniaturize, so small brains have fewer neurons and neural connections, suggesting that tiny brains must have limitations in what they are able to process. Because webs are essentially an archive of subtle decisions, web-building behavior is a classic example of where to find trade-offs associated with small brain size. Adult spiders might be several orders of magnitude larger than young spiders of the same species: So do spiderlings simply build really bad webs? No. An innate behavior, the architecture of a first orb web built by newly hatched spiderlings replicates the adult web.

SMALL BODIES

We are often taken aback by the intricacy of webs. That something so small could create something so complex seems incredible. Then consider that these same structures are built by spiderlings a fraction of the size of adults—some almost invisible to the naked eye. We should be truly amazed. Although web-building is largely driven by a genetic template and requires no learning, building a web requires excellent coordination, the ability to detect the environmental characteristics in which it will be built, and is characterized by a surprising level of plasticity, including adapting to the available space. Adult female *Leucauge argyra* normally construct horizontal orb webs with a diameter of about 36 in (90 cm). However, if confined to a small space in which to build, they make alterations to adapt the web to the space available.

This includes almost 20 key modifications to web architecture, such as smaller spacing between the spirals of the orb webs and less radii emanating from the hub. When building in confined spaces, spiderlings—weighing in at about one hundredth of the weight of the adults—make the same changes with the same level of precision to their webs as the adults.

SMALL BRAINS

All studies estimating the number of neurons in spiderlings compared with adults suggest that neurons are smaller in spiderlings, but they are not necessarily reduced in number, which fits with the lack of evidence that complex behavior is compromised. Since neuron size can only reach a minimum diameter of about 2 µm, this means that spiderlings have relatively larger brains than their adult counterparts. In the closely related species *L. mariana,* the brain is, in fact, too large to house within the confines of the cephalothorax: Neurons from the central nervous system overflow into the legs.

→ Its brain may extend into its legs but that doesn't prevent *Leucauge argyra* falling host to a parasitoid wasp that changes its host's web-building behavior. Rather than its usual rounded orb, an infected spider will create a robust platform in which the wasp larva will build a cocoon for metamorphosis—once it has killed off the host.

VENOMS

The reality of venom

Spiders are not poisonous, but they are venomous: Poisons are ingested, while venoms are injected. Most spiders have venoms, although less than 0.5 percent of spiders have venoms that are toxic to humans. Venoms are widely used among a diverse array of animals but, as may be expected, spider venoms are breathtaking in their chemical diversity and complexity. Along with silk, venom is probably the other main contributor to spiders' success.

↓ Viewed here are the spider's chelicerae, in this case with three teeth, which may be used to grind prey, and the two prominent fangs. Within each fang lies a duct that carries venom from glands in the cephalothorax.

It is thought that the number of existing venom-using spider species (including those not yet formally described) is most likely larger than that of all other venom-using species—from sea anemones to snakes—put together. All wandering spiders rely primarily on venom to subdue prey. Web-building spiders (apart from uloborids) use a variety of tactics in deploying venom.

These include a "silk first, inject venom later" strategy, while others do the opposite, or they apply silk and venom simultaneously or even incorporate venom into the silk.

COMPLEX COCKTAIL

Spider venom is produced in a pair of venom glands surrounded by muscles that can control the release of venom in each gland independently. Each gland has a duct that leads to the chelicerae and fang on its respective side of the head. The venom is then injected through a tiny pore at the tip of the fang, like a syringe. Spider venom systems (glands and ducts) have different sections, allowing them to produce and secrete different venom mixtures, most of which are neurotoxic or act on the nervous system of the recipient. Spider venoms can contain different molecules used as a toolbox to chemically attack their prey. Evidence to date suggests that venoms exhibit variation according to the population or geographical location within a given species and can change according to developmental stage. Male and female spiders also typically have dramatically different chemical profiles to their venoms, which reflect differences in their life history. For example, in species where females hide in burrows and rarely expose themselves to predation risk, it is the males that leave their natal burrows in search of mates that tend to have venoms consisting of chemicals that both subdue prey and that defend against predators.

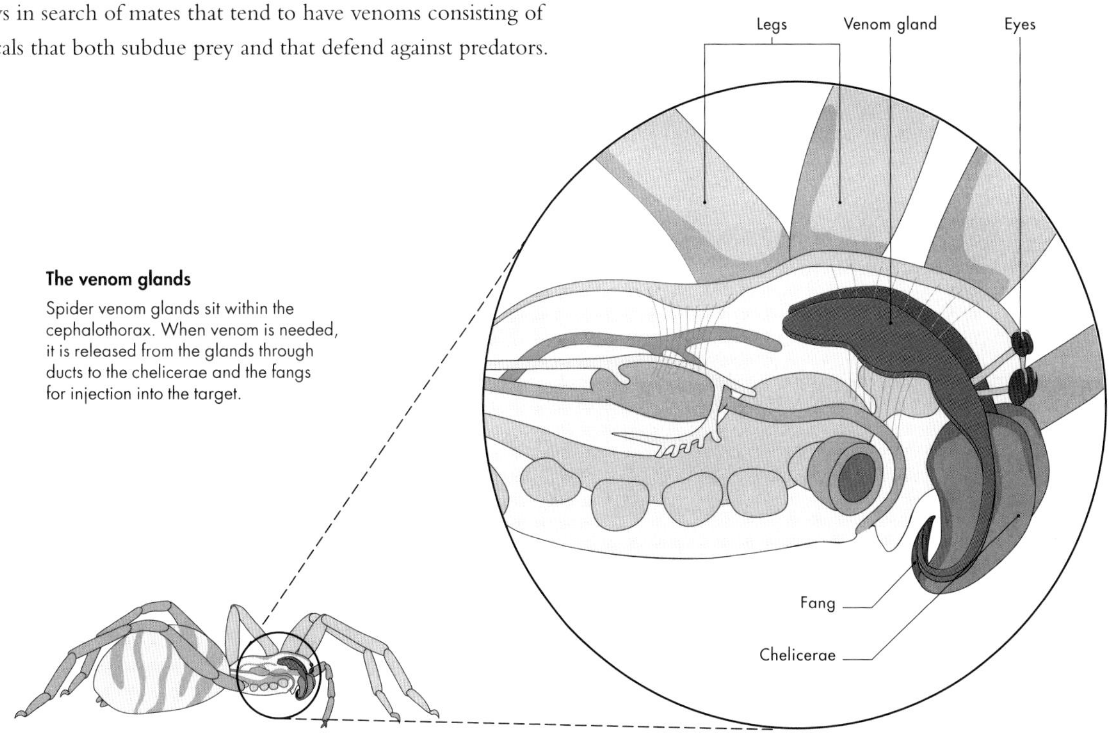

The venom glands

Spider venom glands sit within the cephalothorax. When venom is needed, it is released from the glands through ducts to the chelicerae and the fangs for injection into the target.

Legs   Venom gland   Eyes

Fang

Chelicerae

VENOMLESS SPIDERS

Of 132 described families of spiders, one, the Uloboridae, is venomless. Having lost their venom glands, uloborids cannot rapidly immobilize their prey. Instead, they invest vast amounts of time, silk, and energy extensively wrapping their prey, which can involve over a hundred meters of silk and over 28,000 wrapping movements!

Once the compressed silk bundle is ready, the spiders cover it in potent digestive fluid. Unlike our digestive enzymes, which are strongly acidic (with a pH of 1–3), uloborid digestive fluid is alkaline (pH 10). These enzymes disintegrate any membranous material in the prey, disarticulating legs and other appendages, and further allowing the enzymes to penetrate the inside of the prey, liquefying the innards for consumption. This raises the problematic issue that the membranes of the spiders themselves may be at risk of self-digesting if exposed to their own fluids. For example, *Philoponella vicina* actively move their legs out of the way as they feed, reducing exposure of the membranes articulating their own legs to the digestive fluid. Additionally, they use both thick and thin silk to wrap prey, and the digestive fluid rapidly dissolves the more plentiful thin silk for reuptake. This leaves the prey solely wrapped in the thicker fibers and reduces the metabolic cost of wrapping.

↑ Uloborids are orb-weavers and comprise the only spider family that has lost its venom glands. Instead of using venom, they crush their prey in copious amounts of silk.

→ Spiders whose venom is toxic to humans, like this Sydney funnel web, *Atrax robustus* (see page 122), can be "milked" to collect their venom to produce antivenom. Spiders are now also milked to collect venom for medicinal research.

POSITIVE POTENTIAL

Because spider venom has primarily evolved to subdue or kill their prey—typically insects— the insecticidal properties of their venoms are of considerable interest for pest management. However, spiders only produce an infinitesimally small amount of venom with each bite. These amounts have been insufficient to study spider venom in any detail: Until recently, techniques required large volumes for chemical analysis. Much like DNA sequencing no longer needs puddles of blood, and "touch DNA" can now be sequenced, advances in chemical and molecular techniques now permit us to explore the insecticidal and biomedical properties of spider venom. To date, the venoms of few species aside from those toxic to humans (primarily researched to develop antivenoms) have been characterized in detail. So, with tens of thousands of species yet to explore, the potential discoveries awaiting us are immense.

An arsenal of venoms

Venoms are undoubtedly useful to spiders, but they are metabolically costly to produce and may take time to replenish, so venoms are used sparingly. Spiders adjust venom injected into prey depending on the amount of resistance, thus optimizing their use of venom: A fighter will be injected with more venom than a prey that has already weakened or is small. Spiders also optimize their venom in other ways: for example, by targeting the injection site for maximum effect or by not deploying venom at all (so-called "dry bites") if the prey is deemed sufficiently harmless.

Another way of considering the cost of venom is to look at the evolution of venom glands. Older lineages of spiders tend to be larger, ground–dwelling animals with small venom glands. It is probably no coincidence that the venom glands in newer araneomorph lineages are so large that they can take up a significant portion of the entire head and yet have smaller body sizes. Having sizable deposits of venom seems to allow these spiders to catch prey similar in size to larger spiders without the energy investment of a large body; simultaneously, a smaller body has enabled these spiders to capitalize on flying prey by being small and light enough to hang in aerial webs.

ACTIVE INGREDIENTS

In addition to minerals like sodium and potassium, spider venoms are composed of substances falling into four main groups, each of which has hundreds or thousands of different types of molecules used within the venom. Each group is characterized by the weight of the main compounds that are the primary active ingredients in the venom. In almost all cases, the active ingredients act by blocking or disrupting the function of receptors or ion channels on the cell membrane of neurons in the nervous system, breaking down normal

physiological processes and eventually leading to death. Small nitrogen-based molecules with low molecular mass are especially used by araneid spiders, including several that, in humans, are used as neurotransmitters involved in cell-to-cell communication between neurons, such as serotonin and octopamine. Comprised of long folding chains of amino acid molecules that bestow a high molecular mass, proteins act as active ingredients in a few spider groups. This is particularly the case with spitting spiders and theridiids, including the widow spiders famously toxic to humans. While large proteins also exist in the venoms of other spiders, to date their role is unknown.

The evolution of chelicerae
The cheliceare of the major spider lineages. The Mesothelae and Mygalomorphae have chelicerae roughly parallel to each other, while araneomorph chelicerae face each other and move in a scissorlike fashion.

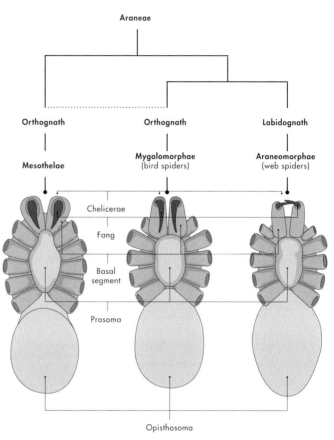

← Brazilian wandering spiders have venom that can be dangerous to humans. Unusually, their venom may be adapted to target small vertebrates rather than insects or other arthropods, which may explain its toxicity to humans.

Most other spiders have medium-weight (molecular weight below 10 kDa, where kilodaltons (kDa) is a measure of molecular mass) peptide-rich venoms, which are comprised of short chains of amino acids. Two major groups of peptides with especially strong insecticidal properties have been identified: fast-acting peptides that are rich in the amino acid cysteine, and antimicrobial peptides that probably play a dual role in preventing microbial infection of the venom gland and aid in paralysis, and perhaps even in digestion.

BENEFICIAL APPLICATIONS

Over the past decade, scientists have been exploring spider venoms for potential application in medicine and in agriculture. The cysteine-rich peptides in spider venoms, in particular, have strong insecticidal effects, and so could provide novel tools for pest control should the correct components be synthetically produced. Destroying over

20 percent of all crops, insects cause almost $500 billion of damage to global agriculture every year. This is a strong incentive to manage insect populations in agricultural areas. However, traditional chemical insecticides are harmful to the environment, and insects rapidly build up resistance to them, meaning that other solutions are needed. Bioinsecticides—which are not harmful to the environment—are particularly needed to sustain the food needs of an increasing global population. Spider venoms have already proved fruitful in this area, with the recent commercialization of the first spider-venom-derived bioinsecticide, with many others still being actively investigated. One potential side effect of these bioinsecticides is nonselectivity, in that all insects (harmful and beneficial) might be killed. However, testing shows that some peptides are highly selective, killing only certain species while leaving others unharmed, paving the way for targeted bioinsecticide development.

←← Widow spiders in the genus *Latrodectus* (see pages 54 and 120) are possibly the most well-known spiders with venom that is toxic to humans.

← The venom of spitting spiders (see page 128) is made up of a complex array of medium-weight proteins and typically cysteine-rich peptides, with glycine-rich peptides possibly being involved in the adhesion properties of their spit.

→ Pest insects, such as this Colorado beetle (*Leptinotarsa decemlineata*), which eats potato leaves, cause massive agricultural damage. As natural insecticides, spider venoms could potentially be used to design target-specific pesticides that are selective to certain pest insects.

MEDICINAL INTEREST

Many cysteine-rich spider venom peptides have very specific and targeted effects on channels modulating ion movement in and out of cells. Because of this, and because they are very potent and act in tiny amounts, these are of considerable medicinal interest. For example, spider venoms often contain compounds that selectively target voltage-gated sodium and voltage-gated potassium channels. These channels are crucial to the function of the nervous system and hence to life itself. Venoms are currently being explored as analgesics for a large range of pain, from inflammation to cancer pain. They are also being investigated as treatments for rheumatoid arthritis, stroke and other diseases involving the nervous system, severe epilepsy, cognitive decline, and several types of cancer. These and other studies involving many additional diseases are showing great promise.

There is growing concern with antibiotic resistance. The World Health Organization (WHO) now considers this to be one of the biggest threats to global health. Recently, studies of spider antimicrobial peptides as potential alternatives to antibiotics have been conducted. Although this work is still in its infancy, once again, there may be some promising candidates. Given the range of potential medical applications already discovered from the venom of the very few species (less than 0.3 percent) for which the venom has been characterized, it is likely that with improved technology and ability to characterize the venoms of more species, this list will increase dramatically.

Venom specialization

In keeping with the notion that venom is costly and its use is optimized, spiders that specialize on specific prey types often have venoms that are especially effective against that particular prey type. This also has the advantage in that it simplifies the composition of the venom.

Prey types that are often targeted by specialist spider predators include ants, termites, moths, woodlice, and other spiders, with ants being particularly dangerous prey to take on. Ants have the nasty ability of being able to bite back, and this is typically lethal. Because of their ability to defend themselves against spiders, most spiders don't hunt ants, and they tend to actively avoid them. However, if a spider can somehow overcome the defenses of ants, it has a huge repository of untapped food with little competition. Ants are often quite large, some significantly larger than the spiders that hunt them, so we might expect ant-eating species to have large venom glands housing enough volume of venom to paralyze such large prey. This is not the case.

ANT-EATING VENOM

While similar effects have been seen in termite-eating, woodlice-eating, and spider-eating spiders, the venom of ant-eating species is best understood. Multiple species of ant-eating *Zodarion* spiders coexist in the same habitat on the Iberian Peninsula. However, each *Zodarion* species seems to target a specific genus of ant

or—being even more particular about preferred prey—a given species, with venoms that target the preferred prey. Unsurprisingly, the venoms of the different zodariid species also differ dramatically, despite their close phylogenetic relatedness. This suggests that prey specialization and venom specialization have coevolved to produce niche differentiation between these coexisting species and preventing intraguild competition for the same food sources. All of this is topped off by behavioral specialization of the preferred prey type without harming themselves in the process.

↑ *Amyciaea forticeps* is a crab spider that mimics weaver ants, on which it sometimes preys. It uses specialized tactics, including hanging from silk threads to bite an ant on the head and waiting for the ant to become paralyzed before approaching it.

← Capable of mounting a communal attack, ants (especially the highly aggressive weaver ants pictured here), are exceedingly dangerous to most spiders.

PREY SPECIALIZATION EFFECTS

Other than being selectively potent against preferred prey, additional effects of prey specialization on the spider venom system include a reduction in venom gland size compared with species that are generalist predators. This presumably reflects a lower metabolic cost to the production of venom among specialists—probably because, compared with generalist spiders, the venoms of prey specialists are much less diverse. These differences between generalist venoms and specialist venoms are prevalent among the peptide components of the venom.

Understanding the natural history of spiders is not only interesting for those of us who do this, but it also sheds insight into the potential applications of this type of research: If I were a biochemist interested in developing an ant-killing insecticide, I'd be looking at the venoms of zodariid spiders.

→ Vertebrate prey, such as this frog, are rarely on the menu. However, if a spider *does* catch a frog, bird, lizard, or other vertebrate, and can subdue the prey with venom, it will make a substantial meal that will likely last it for weeks, if not longer.

Misconceptions

Although the fangs of most spiders can't penetrate human skin and we are clearly not their prey type, and most spiders are reclusive and live away from human habitation, spiders still get a lot of bad press. About half of all media coverage has factual errors, but these are less problematic than the frequent use of blatant "fake news" and sensationalist wording aimed to instill horror and alarm.

One of the most common factual mistakes is describing spiders as "bugs" (or insects) that can sting (using a non-oral, specialized part of the body such a bee stinger to inject venom), as well as naming species incorrectly. More troublesome is wording aimed at instilling dread in a general public and stoking high levels of a largely unfounded fear of spiders. This becomes very clear when looking at the especially high prevalence of arachnophobia (fear of spiders) in areas with no dangerous spider species. Typical words used in media articles include "deadly," "creepy-crawly," "fear," "huge," "slayer," "flesh," "horror," "nightmare," and "terror." Of course, there is also ethically written, factually correct, and even positive press about spiders, but these stories are not typical.

← Numerous B-grade films, with B-grade titles such as *Eight-legged Freaks*, contribute to the mistaken notion that spiders are to be feared.

↘ Large, harmless house spiders, such as this *Tegenaria duellica* cobweb spider, may surprise bath-takers. Spare a thought, though, for the spider—it is most likely a male that, while desperately searching for a mate, fell into the bath, only to be trapped there. Perhaps take him outside so he may fulfil his mission of finding a mate.

SCARE TACTICS

Here are two headlines that a quick Google of the word "spider" and "news" returned, followed by my rewritten headline not aimed to sensationalize: "Man finds terrifying huge nest of Huntsmen spiders lurking in his garage." What this very alliterative headline actually says is: "Man found spider nest in his garage." Hardly going to become the newest online meme, is it? Written that way, it also probably wouldn't heighten a fear of spiders in the public. Or take another one: "Man hoping to turn into Spider-Man after horrific bite in night left him in agony;" this could be translated to "Man takes humorous outlook on sore bite." Who hasn't used humor as an antidote to pain?

Despite the size discrepancy between spiders and humans being solidly in our favor, arachnophobia, present in an estimated 6–11 percent of the population, is probably the most common of animal phobias and is utterly disproportionate to the risk that spiders pose to humans. On the basis that potential harm from spider bites is low compared with, say, being lethally attacked by a bear or lion, a fear of spiders makes little evolutionary sense. Several recent studies have now concluded that arachnophobia most likely originates from a fear of scorpions (and less than 25 percent of these species pose a risk), which has been generalized to include other chelicerates, such as spiders, that are far more likely to be encountered by most people.

THREATENED SPIDERS
....

While in many cases, conservation measures that protect spiders are achieved indirectly through habitat protection (such as nature reserves), in some instances, direct measures to protect spiders have been undertaken. A small number of spider species are protected by law in their respective countries. For example, in the USA, two spiders, the tooth cave spider *Neoleptoneta myopica* and the spruce-fir moss spider *Microhexura montivaga* are listed in the Endangered Species Act, and in Britain, the fen raft spider *Dolomedes plantarius* and the ladybird spider *Eresus sandaliatus* are protected. However, these numbers are stark compared with the scale of the problem. For example, it is estimated that 16 percent of Britain's roughly 650 spider species are under severe threat, while in New Zealand estimates are that about 30 percent of species for which there is sufficient data for assessment (at about 600 species, this is estimated to be less than half of the total number) are under threat, and these are almost all found only in New Zealand.

However, there are signs of change. Caves have been closed to the public in New Zealand to afford the cave spiders *Spelungula cavernicola* protection. In Europe, spiders are mentioned in the regional red lists of two-thirds of countries, and 19 countries have legislated in some way for spider protection. (With Austria listing a whopping 111 species for legal protection!).

Nevertheless, even with the bias against invertebrates in terms of a legal framework for conservation, more needs to be done: About 10 percent of all European butterflies and dragonflies are listed in the EU Habitats Directive, which establishes proception regimes for habitats or species listed, while only a one of an estimated 4,500 spider species is listed in this EU-wide framework. We need your help! Local arachnology societies (usually volunteers) often inform policy managers in their area as to red-list species (threatened, vulnerable, or endangered species) to be considered, and maintain databases that help policy makers manage habitat to promote spider biodiversity. Become a volunteer, and help our spiders.

→ ↘   *Eresus sandaliatus* (top) and *Dolomedes plantarius* (bottom) are the only protected spider species in Britain, although it is almost certain that there are many other endangered species.

3	Extinct
60	Critically Endangered
89	Endangered
61	Vulnerable
17	Near Threatened
89	Least Concern
46	Data Deficient

Extinction risk

The International Union for Conservation of Nature (IUCN) categorizes species according to their vulnerability. This chart shows the classifications for spiders globally. The IUCN asseses species on an ongoing basis, and this chart is correct at the time of writing. Note how few species are on this list (fewer than 1 percent) compared with the 51,500 or so currently described spider species worldwide.

DREADED WIDOWS

Widow spiders (see pages 54 and 120) are infamous for the effects of their venom on humans, yet less than 50 percent of people bitten show symptoms, and less than 2 percent have severe symptoms (no deaths) after a *Latrodectus* bite. That is not to say it shouldn't be taken seriously. But it does put things into some perspective viewed next to an upper estimated annual global death toll from spiders at about five against, say, the frequent and considerable risk of anaphylactic shock from bee stings, a small number of which result in death. Annually, about 70 people die of bee-sting-related anaphylactic shock in the United States. Extrapolating to our current (2022) population of 8 billion people, this means that every year around 1,700 people die of anaphylaxis around the world, or one in 5 million people. Less than one person per billion dies due to a spider bite. So, although bee-related deaths are exceedingly rare, they are still more than 300 times more likely than spider-related deaths.

NECROTIC FINGER-POINTING

Another commonly lambasted group are the white-tailed spiders in the genus *Lampona* in Australia and New Zealand. Media accounts are rife with stories of the severe skin tissue death (dermonecrosis) caused by *Lampona* bites. This turns out to be dramatically exaggerated, with bites typically resulting in some redness. Necrotic lesions attributed to *Lampona* bites are most likely due to another cause. Similarly, in the United States, misidentification of any number of random benign brown spiders leads to the infamy experienced by the brown recluse and other *Loxosceles* (see page 126) spiders, the bites of which can sometimes cause necrotic lesions. However, lesions attributed by medics as recluse bites are also frequently mistaken diagnoses. Necrotic lesions can also be caused by many other things, from bedsores to diabetic ulcers, as well as by bites from several types of arthropods, including ticks, indicating possible Lyme disease. Clear cases of misdiagnoses are those made in areas where recluse spiders are simply not present (but Lyme-disease-carrying ticks might be present). This occurs in part because of their continued poor reputation through unsubstantiated news reports of alleged recluse bites or sightings outside its habitat range.

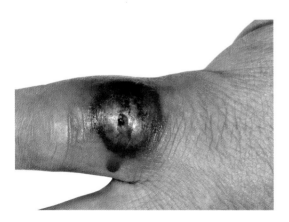

← Necrotic effects of spider bites can look nasty and be very painful, but don't be too hasty to assume the lesion was due to a spider—numerous more likely causes for these lesions abound.

←← Widow spiders are typically recognized by their shiny, bulbous abdomen, although many other spiders share these features. Female widow spiders often have red or orange warning markings on their abdomen, which can make them rather stunning.

CONSERVATION METHODS

While clinical studies in which there is a verified spider bite have clearly demonstrated that only a very small percentage of spider bites cause serious effects, some spiders *can* cause severe envenomation in humans, although the worst affected seem to be young children and the elderly. Also, regardless of whether their venoms negatively affect humans, a small proportion of spiders do have large, stiff fangs capable of penetrating human flesh, and they may inflict a painful defensive bite. Many of these species, such as tarantulas, are also kept as pets. Even if their venoms may not be particularly (or at all) harmful to humans, being bitten by a tarantula would

probably hurt as much as an injection at the dentist, since their fangs are about the size of a large syringe. Generally speaking, it is probably best to handle spiders with care, remembering that for them, we are huge, dangerous giants, so they may be liable to give a defensive bite.

A tolerant public is important if we hope to gain traction in international and national conservation measures for spiders, which are particularly underrepresented in conservation policies. A quick scan of the first section of this chapter should provide us with ample reasons to keep our spider populations healthy. Concerns about spider conservation being low

← Tarantulas are the most common spiders kept in captivity, often as pets. Unfortunately, this has led to many species becoming at risk of extinction.

→ Another way to appreciate spiders is through ecotourism that focuses not just on typical large animals, such as birds and mammals, but on smaller wildlife, like spiders. Often remaining quite still, they can make for excellent photographic opportunities.

on the policy agenda are highlighted by the fact that spiders have become popular as easily maintained pets, thought of by some as "cool." This has led to a large global trade for certain spiders that have become popular as pets, particularly among long-lived tarantulas, of which over 400 species—most of them local only to their region and thus very vulnerable to extinction— are currently "in trade." The International Union for Conservation of Nature (IUCN) monitors global loss of species, listing them as threatened, vulnerable, or near extinction (see page 112), but this list is biased toward more easily observable vertebrate species. There is a general lack of knowledge and incentive to list the

threats to spiders, which means that fewer than 300 spider species are on the IUCN list of threatened species. Based on habitat loss alone, this number should more likely approach the tens of thousands. An even smaller number is listed by the Convention on International Trade in Endangered Species of Wild Fauna and Flora (CITES), which monitors illegal animal trafficking. Fewer than 0.1 percent of spider species are listed by CITES, so the active and extensive spider trade is largely unmonitored.

← The coloring of the this male red-headed mouse spider (*Missulena occatoria*) may be a warning signal about the potency of its venom, which can be dangerous to humans, although bites are incredibly rare. The larger females, in contrast, are typically brown and rarely leave their burrows.

→ Their typically placid nature, combined with often spectacular colors and ease of maintenance makes many tarantulas, like this *Poecilotheria metallica*, or Gooty sapphire ornamental, highly desirable as pets. Unfortunately, this has led them to become critically endangered in nature.

HANDLE WITH CARE

Spider venom experts have concluded that some species in the following genera can cause severe envenomation in humans: Australian funnel web spiders (family Hexathelidae) in the genera *Atrax*, *Hadronyche*, and *Illawarra*; recluse spiders and sand spiders from the globally distributed family Sicariidae in the genera *Hexophthalma*, *Loxosceles*, and *Sicarius*; *Poecilotheria* tarantulas from India and Sri Lanka (family Theraphosidae); *Missulena* mouse spiders found in Chile and Australia (family Actinopodidae); *Phoneutria* or "Brazilian wandering spiders," found from central to South America (family Ctenidae); and species in the widely distributed widow spider genus *Latrodectus* (family Theridiidae). To reduce sensationalist and hyped-up media, evidence-based information on the effects of spider venoms on humans is needed.

You can help: If you have actually *seen* a spider bite you, please pick it up, take it to your friendly neighborhood arachnologist for identification (many universities have an arachnologist, and the spider may be able to be mailed) along with a description or photo of the bite itself. If you are unlucky enough to have any painful symptoms, go to the doctor with a photo of the spider, explaining that it has been sent to an expert for identification. Medical doctors are simply not trained to identify spiders.

Mediterranean black widow spider

Chemical defender

SCIENTIFIC NAME	*Latrodectus tredecimguttatus*
FAMILY	Theridiidae
BODY LENGTH	Females ⅓–⅔ in (7–17 mm), males ⅛– ⅓ in (4–7 mm)
NOTABLE ANATOMY	Has 13 yellow to red spots on the shiny black abdomen
MEMORABLE FEATURE	Painful bite to humans, which rarely can be fatal

One of less than 35 species of *Latrodectus*, clinically the most significant group of spiders worldwide, the venom of this European or Mediterranean black widow spider is possibly the best studied. Venom aids the spider in subduing prey by paralyzing and liquefying it for extraoral digestion. Although humans are not their prey, *Latrodectus* envenomation affects us: The venom contains the vertebrate-specific neurotoxic protein α-latrotoxin, which can cause long-lasting severe pain, muscle spasms, nausea, accelerated heart rate, and, if left untreated, even death on rare occasions.

APOSEMATISM AS AN HONEST SIGNAL

Preferring to live in dark, secluded areas, the much-maligned widow spiders tend to be shy. If prodded, these spiders will typically retreat and hide or flick sticky silk as a mechanical barrier to cover their abdomen. Because venom is metabolically costly to produce, as with many spiders, widows regulate the injected venom depending on need and in defense often only deliver "dry" bites with nonmeasurable amounts of venom. However, if under serious threat, such as if they are squeezed, they can defend themselves with venom. Animals that can repel danger through a painful bite or through chemical defenses such as venoms, often signal this ability through the use of bright, contrasting colors. This has the advantage of being highly salient, and—coupled with it being an honest indicator of potential danger— is easily remembered by animals that have had the misfortune to encounter them. One of the most common "aposematic" signaling patterns is the use of red and black, as found in the adult females of many species of *Latrodectus*. Although little is known about their venom, the fangs of the much smaller males cannot provide a penetrating bite to larger animals. Being typically brown, the lack of male aposematic coloration suggests that any venom use by males is unlikely to be related to defense.

TREATMENT

Widow bites can often be treated with muscle relaxants and opioid-based painkillers. If symptoms persist, antivenom quickly improves the symptoms of lactrodectism. Since α-latrotoxin seems to be present in all species of *Latrodectus*, a single antivenom can be used to treat the bites of any *Latrodectus* and even some related theridiid species, such as false widow spiders in the genus *Steatoda*.

→ Females have 13 spots on the upper side of their abdomen, leading to their scientific name, *tredecimguttatus*, from *tredecim*, for 13, and *guttatus*, for spotted.

ATRAX ROBUSTUS

Sydney funnel web spider

Dangerous shoe-seeker

SCIENTIFIC NAME	*Atrax robustus*
FAMILY	Atracidae
BODY LENGTH	Females 1 ⅜ in (35 mm), males 1 in (25 mm)
NOTABLE ANATOMY	Velvety, glossy brown to black spiders with large fangs
MEMORABLE FEATURE	Extremely venomous to humans

The large mygalomorph funnel web spiders are the world's most dangerous spiders to humans. In contrast to widow spiders, in funnel webs, such as the Sydney funnel web, it is the venom of the slightly smaller male that is especially dangerous—and their fangs are capable of penetrating fingernails. Reaching more than eight years of age, these spiders are long-lived.

LIFESTYLE ENCODED IN VENOM

The composition of the male venom is simpler than that of females, which contains several thousand peptide toxins. Male venom, however, contains large amounts of certain neurotoxic compounds (specifically δ-hexatoxins), which are more likely to be used in defense, instead of to subdue prey. This seems to be due to their lifestyle: Juveniles and adult females only leave their burrows to hunt, while adult males eat little but spend much of their time searching for females to mate with. So males have little need for prey–subduing venom, but are exposed to high risk of vertebrate predation, against which they might need to mount a defense.

In most vertebrates, it seems the primary effect of the venom is pain, but it is lethal to primates unless treated with antivenin. One reason these spiders are especially deadly is that humans encounter them relatively frequently. They are found both in bushland and urbanized areas, such as Sydney. The nocturnally active males seek shelter during the day—and sometimes a shoe will do.

EVOLUTIONARY ORIGINS

Current thinking is that defensive toxins evolve and diverge more slowly than toxins used to subdue prey. Funnel web δ-hexatoxins are evolutionarily conserved, so they are very similar across all 35 species of funnel web spiders and have a shared common ancestry, about 150–200 million years ago. That we and other primates are affected by these toxins is simply bad luck. No primates lived in Australia, where funnel web spiders evolved, until humans arrived about 65,000 years ago. In fact, primates worldwide did not evolve until about 100 million years after the common ancestor of funnel web spiders.

Fang orientation

In orthognath spiders, such as *Atrax robustus* (left), the venom gland (A) is located entirely in the basal segment of the chelicerae (B), but in labidognath spiders, it can extend into the cephalothorax. This also illustrates the difference in the shape of the fangs (C).

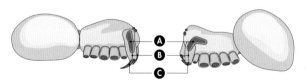

→ When threatened, these spiders display their large fangs by rearing up and prominently extending them.

Ant-eating spider

Ant thief

SCIENTIFIC NAME	*Zodarion germanicum*
FAMILY	Zodariidae
BODY LENGTH	Females ⅛–⅕ in (3–5 mm), males ⅛ in (3–4 mm)
NOTABLE ANATOMY	Bulbous dark-brown abdomen with brown to red cephalothorax and narrow legs
MEMORABLE FEATURE	Specializes in hunting ants

With around 170 species, the genus *Zodarion* is comprised of spiders so specialized at eating ants, they will often starve rather than eat alternative prey. In fact, non-ant prey do not provide the nutrients they require. Furthermore, different species of *Zodarion* specialize at eating different genera or species of ants, possibly to avoid competition. Like its congeners, *Zodarion germanicum* can detect its prey from a distance using the chemical pheromones ants use to signal each other. More than capable of killing spiders, ants are dangerous prey, and *Zodarion* use an arsenal of adaptations to overcome them.

SURVIVING INTERACTIONS

In what might be described as a "wolf in sheep's clothing" foraging strategy, crafty *Z. germanicum* might adopt a low-risk hunting strategy by stealing a dead ant from another *Zodarion* to mask its own odor and avoid alerting other ants of an intruder. In a higher risk maneuver, it might kill an ant at the periphery of the group and then walk through the denser part of the group holding the dead ant in front of it while searching for more ant prey. When making an attack, other behavioral adaptations come into play. Instead of wrapping, or biting and holding prey, *Z. germanicum* performs a quick lunge and bite, followed by a rapid retreat to wait until the effects of the injected venom take hold. This is so effective

that even compared with non-dangerous alternative insect prey, it has increased capture efficiency with ants compared with alternatives.

ANT SPECIALIZATION

In addition to behavioral specialization, *Z. germanicum* is also metabolically and physiologically specialized at feeding on ants. This includes selectively feeding on specific parts of the ant's body (the head and thorax) as a mechanism for achieving a balanced diet, which ultimately leads to faster growth and a longer life span. Crucially, its venom is especially quick at paralyzing ants or even specific ant genera. In fact, *Zodarion* venom takes four times as long to subdue insects that are not ants. Consistent with the idea that prey-subduing venom is strongly selected for and evolves especially fast to keep up with different prey, specific *Zodarion* species have venom that precisely matches the susceptibility of their target ant species. The venom is tuned to specific ants, most likely through species-specific low molecular weight compounds within the venom.

→ *Zodarion germanicum* is thought to be an imperfect mimic of ants, displaying similar coloration and movement patterns

Brown recluse spider

Hidden lurker

SCIENTIFIC NAME	*Loxosceles reclusa*
FAMILY	Sicariidae
BODY LENGTH	⅓–½ in (7–12 mm)
NOTABLE ANATOMY	Typically has a violin-shaped pattern on the cephalothorax and rests with extended legs
MEMORABLE FEATURE	Has venom of medical significance to humans; readily drops limbs when attacked

Of all the spiders whose venom affects humans, it is the brown recluse that is the most mistakenly defamed. An enzyme that is typically found in bacteria and not in spiders (except for *Loxosceles*) can cause "loxoscelism." This condition can produce inflammation, blisters, and lesions, and may be accompanied by fever and nausea.

Most of the approximately 130 species of *Loxosceles* found throughout much of the tropical and temperate world live in areas sparsely inhabited by humans, but the brown recluse is an exception. Commonly found in the United States, where bites attributed to *L. reclusa* are frequent and distributed across the entire country, they are in fact restricted in range to the southern central regions, meaning that only a fraction of bites attributed to them are even possible.

BIOLOGY

The brown recluse lives an inactive life in dry, tight crevices, often in houses, where it typically hunts and sometimes scavenges insects and other spiders. Having a low tolerance for the cold, climate change is expected to expand the range of the species, since it will drift northward, eastward, and westward with increasing temperatures.

→   The brown recluse is sometimes known as the violin or fiddleback spider, owing to the characteristic violin-shaped pattern on the upper surface of the head or cephalothorax.

While the primary role of *Loxosceles* venom is to subdue prey, when threatened they will bite in defense. Having a "reclusive" disposition, humans are not likely to trigger defensive behavior, but seeking dim, narrow spaces, these spiders may, for example, shelter in abandoned crumpled-up clothing, which when put on would certainly seem threatening to the spider and elicit defensive behavior.

LOXOSCELISM OR NOT?

While the venom of males and females appears to be similar, brown recluse bites are usually mild and typically only occur when the spider is squeezed, such as if rolled on in sleep. In exceedingly rare cases, *Loxosceles* bites can result in death due to kidney failure, particularly among children. It is advisable to be aware of the symptoms, although currently no definitive treatment for loxoscelism exists. The symptoms can be confused with more likely candidates, including fungal infections, Lyme disease, cold sores, and especially bacterial staphylococcal or streptococcal infections. The best way to be sure that it is loxoscelism is to collect the spider in the act of biting for future identification; but this is tricky, since pain and symptoms take time to develop. Perhaps the best thing is to be aware of the distribution and habits of your local species of *Loxosceles*, and if you live in an area with a species that often inhabits houses, such as the brown recluse, shake your clothes and check the dark crevices that you might reach into, and then relax—you most likely have an infection.

SCYTODES PALLIDA

Pale spitting spider

Transparent gob-thrower

SCIENTIFIC NAME	Scytodes pallida
FAMILY	Scytodidae
BODY LENGTH	Females ⅛–⅕ in (4–5 mm), males ⅛ in (3–4 mm)
NOTABLE ANATOMY	Very pale with dark stripes across body and legs
MEMORABLE FEATURE	Spits sticky silk to net prey from a distance

Spitting spiders have among the most unusual hunting tactics of all spiders. Despite their poor vision, they are both active hunters and ambush predators, and often attack other spiders. Of course, their spider prey can turn predator.

Spitting spiders reduce the risk of being eaten themselves by spitting a sticky secretion at their prey from several inches away. Ejected from their fangs, the gluey secretion is a glycine-rich type of silk. While it was believed that the sticky spit contains venom, recent work suggests that immobilization is due solely to the adhesive and contractile nature of the silk, which can shrink up to 60 percent after ejection from the fangs. Only after the prey has been spat at will the spider approach it and inject venom with its tiny fangs.

SMALL FANGS, LARGE GLANDS

Scytodes pallida has two large venom glands that produce both venom and glue in different areas, but both are ejected through the same fang duct. In the initial stages of an attack, each of the tiny fangs independently sprays about 12 in (30 cm) of the venom-free contractile silk in a zigzag fashion over the prey in only about $\frac{1}{30000}$, of a second! The force with which the sticky spit contracts compresses the legs of the spider's prey against its body, ensnaring it and reducing mobility sufficiently for the spider to approach and bite it. However, *S. pallida* often scavenges on dead prey and rarely spits in this situation, suggesting that it is metabolically costly.

ARMS RACES

In the Philippines, *S. pallida* specializes in hunting jumping spiders, for which the rapid-fire spit is an effective hunting tactic against these visual predators. In some instances, *Scytodes* builds its sparsely woven web over the jumping spider nest and attacks it as it leaves or enters. Because of this, some jumping spiders actively live near ants—at considerable danger to themselves—because the ant odor deters *Scytodes*. In turn, some spider-eating jumping spiders, such as *Portia labiata*, have evolved strategies to hunt *Scytodes*. Spitting spider mothers carry their egg sac in their chelicerae, rendering their primary long-range weapon useless. Hunting jumping spiders therefore identify and target egg-carrying spiders as prey. In turn, *Scytodes* can detect the odor of nearby *Portia*, and would-be *Scytodes* mothers can modulate the time it takes for their eggs to hatch, shortening incubation time in the presence of *Portia* odor, minimizing predation risk. Incubation time is cut further when the odor of *Portia* emanates from spiders that have recently fed on *Scytodes* compared with flies, illustrating the plasticity of key life history parameters and the sensitivity of *Scytodes*' sense of smell.

→ After this female's eggs hatch, the juveniles may remain in their mother's web for an extended period, with the mother taking prey to her young and sometimes feeding alongside them.

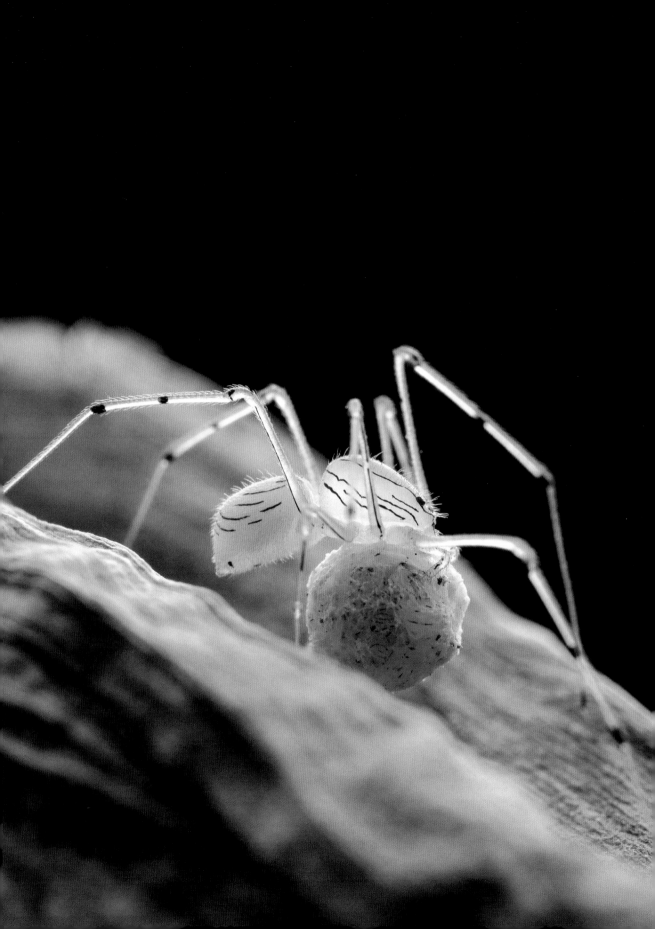

SPIDERS AS
PREDATORS

The predatory role of spiders in ecosystems

The habitats of spider species are many and varied. Some inhabit flowering inland gardens or lush rain forests, some thrive in desert locales, some make their webs near flowing rivers or bodies of water, and some make their webs in farmland. All spiders become an integral part of their ecosystems and adapt their predation styles to suit their needs. An increasing number of species are being affected by human intervention of the environment.

A stream flanked by reeds and extensively branching trees bubbles through a park. Beyond are shrubs and more trees—some large, providing dappled light to a range of low-lying flowering plants below. The ground is similarly diverse, with mulch giving way to drier soils and eventually to rocks and pebbles approaching the river's edge. In this hypothetical garden, there exists a microcosm of life. The larval stages of many insects hungrily feed in the stream, only to emerge as adult mayflies, caddisflies, dragonflies, or damselflies flitting above the water looking for a mate or prey, while others, such as water striders, hunt on the water's surface. More terrestrially bound insects roam the soils, pebbles, trees, and shrubs: ants, beetles, bugs. The flowering plants attract bees, wasps, flies, butterflies, and moths. Insect life abounds, attracting predators: spiders.

→ Bromeliads perched high in the trees make for excellent habitat for spiders. The plants serve as repositories of water for a diversity of life because their structure allows for moisture to pool, creating miniature lakes and providing spiders with plentiful food and an excellent place from which to hide from their own predators.

HABITAT ADAPTATIONS

Some spiders, particularly araneid (orb-weaving) spiders, spin large webs between branching trees, catching flying insects in their nets. Others, such as wolf spiders, patrol the soils and pebbles in search for earthbound prey. Some, like tetragnathid spiders, will live by the water's edge, with parts of the web occasionally dangling into the water, ready to ensnare adults of the aquatic insects provided by the stream. Yet others, such as crab and lynx spiders, will make their home amid the flowers, ready to ambush a pollinating insect as it searches for nectar. Even within these microhabitats, there could be a degree of further subdivision. For example, some species might favor certain plant species with a given architecture, and will rarely be found elsewhere. This scenario describes how many different species of spiders might live in a small area and reduce competition for the prey available. Habitats, in turn, differ dramatically. Desert habitats might be dominated by stones, succulent plants such as cacti and agave, and low-lying shrubs without a stream to be heard. Rain forests will be dripping with water, mosses, bromeliads, and a plethora of foliage from the

ground high into the sky. Alpine systems, with steep, rocky, rivers gushing down the mountainside, will change especially dramatically with the seasons, potentially being snowbound in winter to having low-to-the-ground flowering shrubs emerging between the rocks in the spring. Each of these alternate scenarios will attract a different suite of insects and, in turn, a different suite of spiders to eat them.

AGRICULTURAL PREY

All the ecosystems described above are largely undisturbed by human intervention, but much of our planet is not this way. A significant portion of our terrestrial area is used to produce food. Agriculture is often very limited in its biodiversity, with a single crop grown over vast tracts of land. This poses issues for the diversity of the insect life that it can sustain, yet some life still

↖ Spiders living in arid or desert areas must contend with a relative scarcity of prey and usually extremely high daytime temperatures, often prompting them to become nocturnal.

↑ Fishing spiders can often be found by the edges of streams, from which they hunt insects and even small fish.

↗ Some spiders, such as some ground-dwelling wolf spiders, are remarkably resilient and even able to survive on agricultural land that is dominated by a single plant type (known as monoculture).

exists. Notably, these are typically insects that feed on those crop plants, and so are considered pests, since they directly compete with humans for the food provided by those crops. The economic impact of insect pests on agricultural yield is immense, so we develop methods to kill them. These are typically "broad spectrum" chemical-based sprays that are not selective in what they kill, having detrimental effects equally on pest aphids as on key pollinators for many of our crops (without which we could not survive). Spiders also appear to be affected by these chemicals. However, ground-based spiders less exposed to sprays targeted at pests of fruit trees seem to do fairly well in these systems, eating the ground-based insects that are also less exposed to the spray. This suggests that

harnessing spiders to act as biocontrol of pest insects that eat the crops may be a good idea; but the truth is that, in most systems, particularly in intensive monocultures, this is unrealistic. However, in diverse forms of agriculture, such as in orchards, spiders are often actively used as biocontrol agents that cause no environmental damage.

Whether it is in natural habitat or properly managed agricultural habitat, spiders have a significant role to play in controlling insect numbers. Understanding spider behavior, dietary preferences, habitat niche, and venoms will allow us to consider these characteristics to better plan our agricultural practices and harness their freely available spider power.

Active foragers, sit-and-wait predators, and a useful arsenal of tricks

→ Tarantulas often burrow in leaf litter and soil, but may also seek refuge in tree cavities that provide them protection from predators by hiding them from view. Here, they can lie in wait to ambush prey.

↙ Tarantulas are sit-and-wait predators, so if you see one wandering through the leaf litter, it is most likely a male searching for a mate.

The widespread idea that spiders are simplistic "automatons" that simply respond in a fixed manner to any given stimulus, such as rushing to attack the source of vibrations on a web, has now been firmly debunked. Spiders respond with extreme subtlety to external sources of information and show variation between species, within species, and between individuals in terms of how they respond to stimuli at any given time.

Aside from sex-specific differences, individual responses depend on hunger level, age, and a host of environmental variables, such as wind conditions and ambient humidity. Given this diversity, it is unsurprising that there are many different foraging techniques adopted by spiders. These can be roughly divided into two camps: active predators that walk around and look for prey in the absence of a web (wandering spiders); and sit-and-wait (ambush) predators that remain in one fixed place for long periods of time, attacking when prey are nearby. In the latter group, we find the classic web-hunters, which stay in or near their web and attack prey caught in the web. This group also contains burrowing spiders, such as most tarantulas, some wolf spiders, and trapdoor spiders, which ambush prey passing near the entrance to the burrow; and other ambush hunters, such as crab spiders and many huntsman spiders, which sit in place quietly on a flower or tree trunk and are often well camouflaged.

COLLECTIVE STRATEGIES

While it is convenient to envision ambush spiders and wandering spiders as nonoverlapping categories, things are predictably not so simple. An individual wandering spider may shift between predation from the nest, such as the burrow, and predation away from nests, in the same way as a web-building spider may build its own web, surreptitiously invade the webs of other spiders, and even capture prey away from webs. And that doesn't include spiders that live socially and are collectively involved in prey capture, or those that scavenge on already-dead arthropods. Nor does it include the use of plants as food. While spiders are famous for being an exclusively predatory group of animals, it is becoming increasingly clear that many spiders, from web-builders to wanderers, do also consume plant matter.

Some spiders engage in collective capture of prey, perhaps like some slowed-down version of a miniature lion hunt. It is tempting to conclude that any spiders that aggregate in colonies will participate in cooperative prey capture, but this is seldom the case. Many spiders that aggregate in colonies are composed of territorial individuals that build their own webs, but build them in a cluster. These subsocial spiders benefit from extended maternal care, which includes regurgitating food to provide for the young, like birds do for their chicks. Foraging benefits for these spiders also exist. As the webs are bunched-up, insects bouncing off neighboring webs provide an extra bonus, and although they may not cooperate in prey capture, they may share prey with other members of the group. However, in these spiders, the pampered juveniles do eventually leave the comfort of their natal group to form new aggregations. Because of the added benefits provided by multiple webs, most subsocial and social species are web-building, but some non-web-building crab spiders (for example, *Australomisidia ergandros*, *Australomisidia socialis*, and *Xysticus bimaculatus*) also exhibit extended maternal care and share large prey that are ambushed and subdued more quickly when acting as a group.

→ Among African social velvet spiders, (*Stegodyphus mimosarum*), a small number of spiders cooperate to capture prey, which they then share with the other members of the colony.

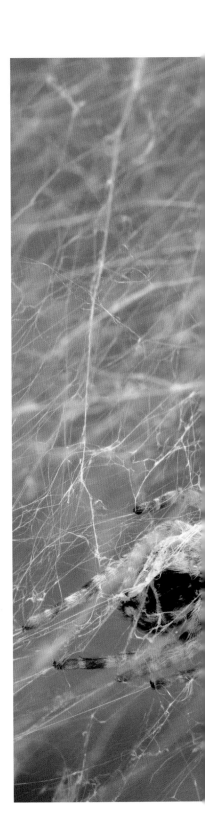

SHARING PREY

A step up in spider social bonds is found among social spiders, especially often among *Stegodyphus* (see page 160) and *Anelosimus* species (see page 88), both of which also contain subsocial species. Here, closely related family members build communal webs. Social spiders evolved from subsocial spiders by the loss of a dispersal stage, whereby instead of juveniles leaving the natal area to form new groups, they simply remain put, reducing the risk associated with dispersing to a new area. These spiders often cooperate in prey capture, allowing them to exploit prey that would be impossible for a single individual to kill, such as large prey relative to the spider's own body size. The added strength provided by the communal webbing means that webs can withstand the struggles of exceptionally large prey without breaking, while the vibrations of the struggling prey are detected by multiple individuals. These spiders use teamwork to attack the hapless prey in a coordinated manner. There are numerous aspects to coordination and cooperative feeding. For example, the venom of one spider might not be sufficient to subdue the prey, but that of many can. Similarly, the digestive enzymes of a solitary individual may not be enough to efficiently extract nutrients from the prey, but the cumulative effect of enzymes from multiple individuals increases the ability of any given individual ability to extract nutrients, so that more nutrients are extracted per unit of time by group-feeding spiders than by solitary spiders.

Other tricks in the predatory arsenal used by spiders are highly variable web decorations, in which spiders use silk, known as stabilimenta, to "decorate" their webs, typically near the hub of the web. These decorations are called stabilimenta because they were thought to help stabilize the webs. Although the geometrical structure of stabilimenta may vary between individuals and in the same individual, species-specific general patterns of geometry—such as zigzags, dots or tufts, spiral disks, or X-shaped ("cruciform") structures—are used.

↑ Web decorations, or stabilimenta, come in many shapes and sizes. Here, an *Argiope* has constructed a line. A common alternative is to make another line at right angles, forming an X, or cruciform, shape.

↗ Spiral stabilimenta are also common. This image depicts a stabilimentum reflecting ultraviolet light. This property has led to hypotheses concerning the role of web decorations in attracting or deterring animals capable of seeing UV light.

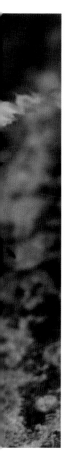

UNUSUAL DECOR

Some spiders decorate their webs with prey remains or detritus. The function of these decorations and of stabilimenta is a hotly debated topic. While silk decorations may in some cases help with web tensioning (and therefore in prey detection), this is unlikely to be true of decorations consisting of detritus and prey carcasses, which may simply be, at least in some instances, food pantries to store prey in times of plenty. Yet, this does not explain why some species attach leafy material to their webs as decorations. It seems likely that in many cases, these non-silk decorations may function as conspicuous decoys that divert predatory attacks away from the spider and on to the distracting artefacts.

Other functions specifically of stabilimenta include thermoregulation. Here, decorations provide shading for the spiders, which, being cold-blooded, are unable to regulate their body temperature internally and are thus reliant on external measures to regulate their temperature. Given the shapes, such as zigzags, of many decorations, this also seems unlikely to be universally true. One thing that spiders are not is shape-shifters. Proposed visual functions have more consistent support. Here the idea is that the decorations influence the interactions between spiders, their predators, and their prey through visual cues, such as being attractive to prey, for example, by being UV-reflective like many flowers or by being aversive to predators, possibly acting as some form of shield.

Specialists and generalists

Most spiders are generalist predators that typically eat insects, and particularly among social spiders, prey can be several times their own size. A small but significant minority of spiders, especially among the wandering spiders, have specialized in eating a given type of prey. Often, but not always, these are prey that are avoided by other predators, including spiders, because they are well defended.

Moths might seem clumsy and good target prey, but they are so scaly that they tend to be slippery and difficult to pin down, but some spiders do attack them. Woodlice are also selected by certain spiders: They have tough external armor and often roll into an impenetrable ball as a defensive tactic. Others, such as ants and spiders, are generally avoided because they are outright dangerous and could easily turn things on their head for the spider looking to make a meal, since it could turn into its meal's meal.

Specialized predators, as we have seen, sometimes have venoms targeted for their efficacy against the specific prey. Similarly, they have particular patterns of predatory behavior that maximize their ability to catch those types of prey. By combining effective venom, morphological adaptations, and effective prey-catching techniques, these spiders have evolved the full range of predatory adaptations required to catch prey. In many cases, this occurs to the detriment of catching alternative prey, so there are good reasons these spiders have specialized, among them the lack of intraguild competitors.

VERTEBRATE REPAST

Notwithstanding their role as predators of insects, some spiders do capture vertebrate prey. This may include the incidental lucky break of a small lizard caught in a web, but other spiders may feed on vertebrate prey more commonly than we have previously assumed. Prey consumed typically include tadpoles and frogs, small fish, small birds, and lizards and geckos. Flashy names like "Goliath birdeater" (the tarantula *Theraphosa blondi*) don't really help root out how common the phenomenon of feeding on small vertebrates is among spiders. However, about 9 percent of spider families do have species that have been relatively frequently seen with vertebrate prey. Although these are still very rare occurrences, they are not as outlandish as they were once thought to be, and vertebrates may constitute up to 5–10 percent of these species' prey (typically closer to 1 percent). Unsurprisingly, spiders that are typically seen feeding on vertebrates tend to be large. The Goliath birdeater, for example, is considered the heaviest of all spiders, with an individual weighing in at a whopping 6 oz (170 g).

↗ Over 90 percent of all instances of predation on vertebrates has been reported in only about 7 percent of all spider families. Unsurprisingly, spiders in these families tend to be large. The Goliath birdeater here is the largest spider of all.

→ Some vertebrates may be ambushed; others, such as this lizard in a *Trichonephila senegalensis* web, may be caught in strong, highly efficient webs.

← The "roly-poly" defense of woodlice is especially effective against most spiders; that is, unless they have specialized means of attacking them (see page 156). This fishing spider is unlikely to succeed.

That spiders rely, in part, on eating nectar from plants is now well established, although not generally known. This behavior is especially prevalent among juveniles that may be too small to easily catch insects. For example, juvenile web-builders, such as the European garden spider *Araneus diadematus*, unintentionally feed on microscopic pollen particles that are stuck to their web when they take down and consume their web to recycle the proteins. While this may not be purposeful behavior, the energy provided by the pollen can double their chances of survival in these early life stages. Many wandering spiders, in contrast, actively seek out pollen by visiting flowers and hunt for nectar at specialized plant pores that excrete nectar away from the flowers themselves (extrafloral nectaries). Some spiders, such as the jumping spider *Evarcha culicivora* (see page 154), actively look for certain plants from which to feed on plant products.

NUTRITION REGULATION

Regardless of their foraging mode, spiders can also actively regulate the protein and lipid nutrients that they consume to compensate for nutritional imbalances. For example, a spider might consume more of a given prey if it contains the nutrient that is currently lacking in its diet and less of one if it is already well provisioned for that nutrient. Spiders deficient in a given nutrient are also able to extract selectively those specific nutrients from a single prey item, while leaving behind the unneeded nutrients. This requirement for nutrient balancing may be the reason why so many spiders are omnivorous, since carbohydrates from plant-derived foods are probably an important nutritional supplement to spiders that normally obtain lipid and protein-rich foods in the form of prey.

↗ Ambush predators, like this crab spider, appear cryptic to the human eye, but may be salient in the UV-capable eyes of their prey—so salient, in fact, as to warrant further inspection.

↓ *Portia* jumping spiders specialize in hunting other spiders, often taking the risk of invading their webs. This strategy can easily turn, and the hunter may become the hunted.

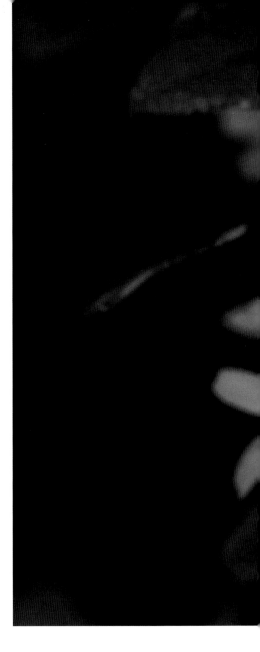

PREY-LURING TACTICS

Prey specialist spiders often employ specific tactics to hunt their preferred prey. One tactic often used by these spiders is aggressive mimicry, where the spider "lures" the prey toward it using deceitful signals. As used by bolas spiders, these signals may mimic the signals given by females of a given animal, such as a moth, attracting males toward them—except that no ready-to-mate female awaits the males that respond. Other spiders may mimic the vibrations of a potential

prey caught in the web of the target spider, drawing the resident spider out to within attacking distance of its hunter.

Several species of the jumping spider *Portia* use this tactic. They specialize in hunting other spiders, and sometimes these are web-building spiders. Regardless of their particular preference, *Portia* has a deceitful trick up its sleeve: Some species that hunt other jumping spiders may tap the leaves where the resident jumping spider is nesting, mimicking the vibrations of a male approaching a female's nest and luring her out, as a kind of reverse-sex bolas spider approach, but in a different sensory modality. Other *Portia* use their palps and legs to pluck a tune on the web of a spider, which lures the resident out slowly toward *Portia*. Too fast an approach could be dangerous to *Portia*, so the tune plucked on the strings is meticulous and can be altered to suit the circumstances and specific species of prey being drawn out.

Spiders as biocontrol of pests and bioindicators for ecosystem health

Spiders being remarkably abundant, having a variety of hunting strategies, and existing just about everywhere, it is natural to consider the role that they play in the control of other insects, especially those that are considered pests of crops or invasive insect species. The use of one organism to control the numbers of another—rather than through chemical controls such as sprays—is known as biocontrol.

Amazingly, there are no studies on how local spiders might suppress numbers of accidently introduced insects anywhere in the world. However, evidence suggests that spiders do have a significant role to play as biocontrol agents of pest species, which are themselves often introduced. This is particularly clear in tropical regions with higher biodiversity, both of spiders and their prey.

In sufficient numbers, spiders can have strong effects, not only on target pest species, but also on secondary pest species, and this often leads to increased crop yield. Stronger effects are apparent when the crop

↗ As generalists, spiders are important predators of insects that cause crop damage, such as caterpillars, or that are vectors for major diseases, such as malaria-carrying mosquitoes.

← Caterpillars and other insect pests can cause immense losses to our food crops.

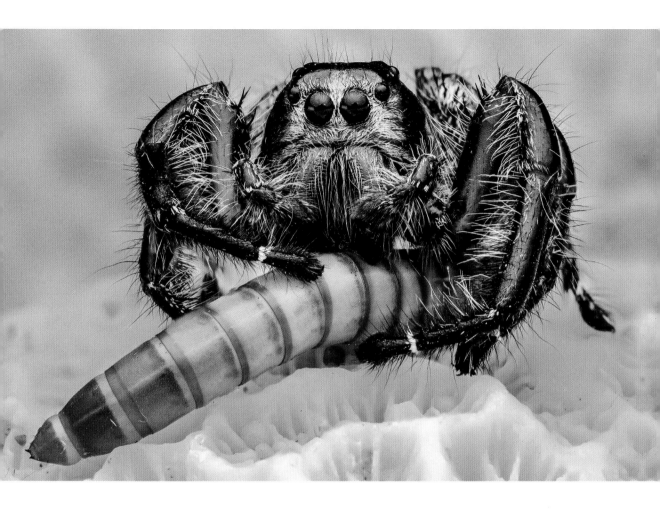

type is either spatially complex (consider corn versus a blade of grass or wheat), perhaps by providing many sheltering opportunities, or when there are a variety of crop types within an area, again providing diversity in spatial architecture. Because insects can destroy almost a quarter of all crops, which leads to huge economic losses, thinking of how we plant our crops (for instance, interspersing different crops rather than planting huge swathes of the same type), as well as considering the natural enemies of the pests of those crops are fruitful avenues to explore. This is especially true in an increasingly environmentally fragile world that can ill afford to overuse chemicals that kill organisms indiscriminately.

ECOSYSTEM MONITORS

Spiders can also help us monitor the fragility or stability of our ecosystems, as well as indicate the success of restoration management practices. For the latter, it may take hundreds of years for secondary regeneration— for example of a forest—to attain the full development of a primary, unaffected, site, and this is somewhat difficult to monitor. However, the assortment of spiders in the regeneration sites may give an indication as to how well the projects are going, as, typically, waves of new colonists of different guilds and species will arrive depending on the habitat types provided by the site. It follows, then, that understanding these waves gives us an eye into the future and to the success of the regeneration plan.

More common is the use of spiders to provide an indication of
the current situation in a given ecosystem. This may be through
the constant monitoring of species abundance and diversity, which
can provide a very good indication, if, for example, a chemical spill
has occurred nearby. These are fairly simple. For example, a 20 yd
(18 m) line (transect) can be implemented at various points along
the site. Someone consistently walks this at a certain interval of
time, accounting for seasonal effects, counting the number and taxa
of animals present along that line. Say that we know that in spring
we can roughly expect a given number of spiders and a given
number of taxa within spiders along those lines; we can again
measure these in the spring following the spill. If there is a
substantial reduction in spider number and diversity, this tells you
that the insects providing them with food are also in poor health.
Spiders are "indicator species" that allow you to infer the health
of the larger ecosystem, because they are major predators of insects.
While spiders are not "apex predators" (the top predator in an
ecosystem), this is a beneficial trait for an indicator species. By the
time an apex predator is seriously affected, the system will have
essentially collapsed. It is the key intermediate predatory guilds
that hold insight to the health of the ecosystem and can allow
us to remediate the problem, such as through regeneration,
before the apex predators are too severely affected.

HEALTH INDICATORS

Naturally, this method can also be used to simply monitor
ecosystem health. Because different spider species select different
areas of a microhabitat in which to live—such as burrowing in soil
or using plants for nest construction or web-building—different
species can also provide us with an idea of how those specific
aspects of the microhabitat are doing. An analogy would be to use
a coarse-grained test for blood typing versus more detailed measures
to obtain a genomic-level understanding of the source of a medical
problem—a more thorough forensic approach. How we fix detected
problems is up to us: Do we use the bandage approach, or do we root
out the problem? The economic implications of the latter should
not be overlooked, but nor should the economic implications of
a bandage failure. Spiders are neither bandages nor ultimate solutions,
but they are a sensible part of the tool kit.

Anthropogenic
point source of
contamination
discharge

Spiders as bioindicators

Spiders are mesopredators, sitting in the middle of the food web. They are predators of both aquatic and terrestrial arthropods, but are predated on by other spiders and vertebrates. Because some spiders feed on insects that spend their larval stages in water, spiders can accumulate the toxins ingested by their prey, making them good bioindicators of environmental health.

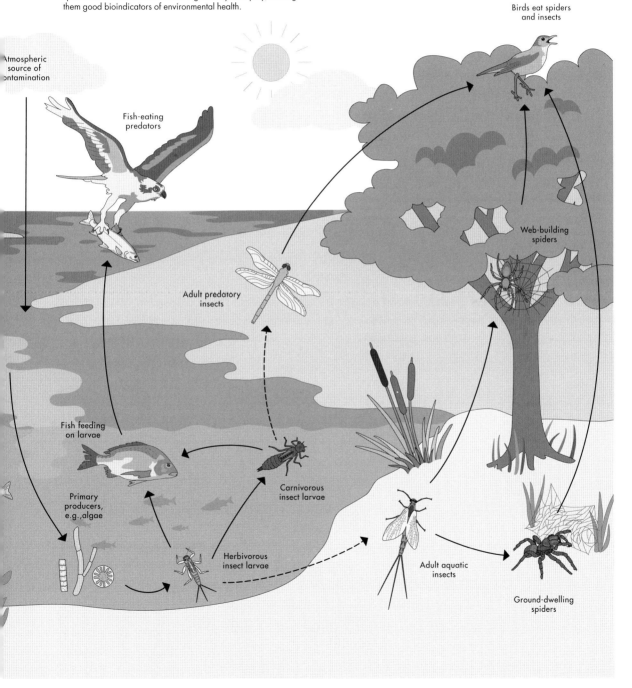

Birds eat spiders
and insects

Atmospheric
source of
contamination

Fish-eating
predators

Web-building
spiders

Adult predatory
insects

Fish feeding
on larvae

Carnivorous
insect larvae

Primary
producers,
e.g.,algae

Herbivorous
insect larvae

Adult aquatic
insects

Ground-dwelling
spiders

BAGHEERA KIPLINGI

Vegetarian spider
Ethical eater

SCIENTIFIC NAME	*Bagheera kiplingi*
FAMILY	Salticidae
BODY LENGTH	⅕–¼ in (5–6 mm)
NOTABLE ANATOMY	Males have iridescent green markings on cephalothorax and abdomen
MEMORABLE FEATURE	Primarily vegetarian

A spider is an unlikely vegetarian, but *Bagheera kiplingi* almost fits the bill. Supplementing its diet with nectar, ant larvae, and nectar-feeding flies, this jumping spider feeds almost entirely on Beltian bodies, the detachable fat and protein-rich leaf tips of *Vachellia* acacia shrubs. *Bagheera* is so dependent on Beltian bodies that it is an obligate resident of *Vachellia* plants, where it lives in areas that are not well patrolled by the resident *Pseudomyrmex* ants. There is such host specificity to the plant that the spider's geographic range is limited by the presence of *Vachellia*.

PLANT MUTUALISMS

Ants can be helpful to plants because they tend to be aggressive and keep herbivorous insects away. Consequently, many plants make an effort to lure ants as bodyguards and keep them around by producing accessible nectar through extrafloral nectaries. This continuous source of food is irresistible to ants, but often is also exploited by spiders, especially wandering spiders that roam to hunt their prey. This includes many species of jumping spiders, where nectarivory may be a common tactic to obtain a meal with less risk of injury than hunting. Nectarivory can increase spider longevity and reproductive output. Importantly, for the tiny spiderlings, nectar may provide much-needed energy that allows them to hunt prey inevitably larger than themselves.

In addition to extrafloral nectar, *Vachellia* species produce nutritious Beltian bodies to keep *Pseudomyrmex* ants nearby. The defense put up by the ants is formidable, and few animals can encroach it. *Bagheera* exploits the mutualism by harvesting the Beltian bodies and extrafloral nectar produced by the acacia without providing defense to the plant. Being able to see ants from a distance, *Bagheera* largely seems to avoid encounters with them—unless craftily stealing a larva being carried by one.

AN UNUSUAL DIET

Depending on location, plant-derived food accounts for between 60 and 90 percent of *Bagheera*'s diet, making this the only near-herbivorous spider known and a rather extreme outlier in a group known for its predatory behavior. As spiders cannot ingest solids, the Beltian bodies must be enzymatically broken down prior to being consumed, which can happen in a matter of minutes. Although this may be an easily available source of food, the spiders appear to need a lot of it to get by: They feed on many Beltian bodies in a single feeding bout, and about 30 Beltian bodies are required to provide the nutrition of a single insect prey.

→ The genus, *Bagheera*, was named for the protective black panther in Rudyard Kipling's *The Jungle Book*, with the species named after its author.

EVARCHA CULICIVORA

Vampire spider

Blood craver

SCIENTIFIC NAME	*Evarcha culicivora*
FAMILY	Salticidae
BODY LENGTH	⅛–⅖ in (3–10 mm)
NOTABLE ANATOMY	Males have bright-red band under forward-facing eyes
MEMORABLE FEATURE	Specializes in hunting the vectors of *Anopheles* (malaria) mosquitoes

Living in the Lake Victoria region of Africa, *Evarcha culicivora* is possibly the pickiest animal on Earth. The media-named "vampire spider" does not feed directly on human blood, but does so indirectly by preying on blood-fed female mosquitoes. In fact, *Evarcha* actively chooses *Anopheles* mosquitoes, which are attracted to feed on human blood and are hence vectors of malaria.

By feeding on blood-fed female *Anopheles* at a time of day when the mosquitoes tend to rest after a blood meal, sexually mature spiders attain a "perfume" that makes them alluring to the opposite sex. This suggests that, unusually, their prey preference may be at least partly driven by sexual selection. As a coup, *E. culicivora* may play a small role in mitigating the transmission of malaria by preventing mosquitoes carrying the parasite from biting and infecting another person.

AN AFFINITY FOR BLOOD

The vampire spider has an approximate hierarchy of preferences, with blood-fed female *Anopheles* at the top, followed by other kinds of local blood-fed female mosquitoes, then non–blood-fed female *Anopheles*, male *Anopheles*, and finally the most common prey type in its habitat: midges. Juveniles even have an *Anopheles*-specific method of hunting, which they don't use for other prey. Odors associated with humans may attract the spiders to houses, where they are likely to encounter the *Anopheles*, but it is their visual decision-making that we understand best. *Anopheles* has a specific resting posture, and *Evarcha* uses this to differentiate it from other mosquitoes. The spider judges how "fat" the abdomen appears as an indication that it is full of blood. To determine sex, it also looks at how feathered the antennae are, as female mosquitoes have barer antennae.

PARADOXICAL PLANTS

Aside from houses, a popular hunting spot is on *Lantana camara* shrubs, where mosquitoes sometimes rest and eat nectar. The spiders also feed on *Lantana*'s nectar, which gives them a nutrition boost that allows them to hunt prey many times their size. Paradoxically, *Evarcha*'s prey preference is no longer expressed when the spider is exposed to the dominant volatile compound of *Lantana*, β-caryophyllene. This is because the plant odors reduce the time *Evarcha* spends visually assessing its prey. The fact that the spider is prone to identification errors of its preferred prey illustrates a trade-off in *Evarcha*'s ability to process information when faced with a diversity of stimuli involving multiple sensory modalities.

→ There may be reason to believe that the vampire spider is one of the most discerning eaters on Earth.

Woodlice-eating spiders

Deadly manipulator

SCIENTIFIC NAME	*Dysdera crocata*
FAMILY	Dysderidae
BODY LENGTH	Females c. ½–c. ⅗ in (11–15 mm), males c. ⅖ in (9–10 mm)
NOTABLE ANATOMY	Has very noticeable and broad chelicerae
MEMORABLE FEATURE	Specializes in hunting woodlice

Woodlice are terrestrial crustaceans (isopods) with a thick carapace, which they use as a shield when they roll into a ball or cling to a surface to avoid attack. Despite being slow-moving, many species have noxious secretions, making them formidable foes. Some spiders in the genus *Dysdera*, the most famous being *Dysdera crocata*, are among the few predators to hunt them.

PINCER, FORK, AND KEY

Species that specialize in catching woodlice have specially adapted chelicerae. Unlike nonspecialist *Dysdera* species, these specialists use one of three main tactics to grasp prey: the pincer, the fork, and the key. Each strategy is associated with a particular mouthpart morphology. Species with elongated chelicerae, like *D. crocata*, use the pincer approach, rapidly penetrating the unprotected underside of a woodlouse with one chelicera before it can roll up and defend itself, while simply holding the armored side to keep the prey in place. If the woodlouse manages to roll into a ball or cling hard, the spider patiently waits, unmoving and ready, until it gets another chance to attack. The fork tactic is used by species

that have chelicerae with a concave upper surface. Here, attacks consist of quicky grabbing the woodlouse with its first pair of legs, slipping the chelicerae under the isopod, and rapidly biting the underside of the woodlouse before it has time to adopt a defensive posture. The key tactic requires flattened chelicerae. Like fitting a key into a lock, the spiders slide one chelicera between the armored segments of the carapace of the woodlouse, inserting its fang to bite—voilà!

WOODLOUSE GRADIENT

Of the *Dysdera* species that largely consume woodlice, there is variation in how much they rely on these prey. However, it is likely that all need to eat at least some woodlice to grow and develop quickly, suggesting a metabolic need for this food source. Furthermore, there is a correlation between the level of modification of the chelicerae and woodlouse specialization, with those that are almost obligate specialists having the most strongly reshaped mouthparts. This is matched by behavior, with species with less modified mouthparts exhibiting markedly less prey preference, and by their ability to extract key nutrients from their prey.

Chelicerae of woodlice spiders

(A) Unmodified chelicerae for more generalized predation; (B) Elongated chelicerae for specialized woodlouse predation using the pincer tactic; and (C) Flattened chelicerae for woodlouse predation using the key tactic.

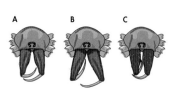

→ *Dysdera crocata*'s fangs, which are capable of penetrating woodlice (or pillbugs) can deliver a harmless, but painful, bite to a human.

St. Andrews cross spider

Web decorator

SCIENTIFIC NAME	*Argiope keyserlingi*
FAMILY	Araneidae
BODY LENGTH	Females ⅖–⅗ in (10–16 mm), males ⅛ in (3–4 mm)
NOTABLE ANATOMY	Females have bright, banded abdomens with two yellow stripes
MEMORABLE FEATURE	Sits with legs in pairs forming an X shape

Many orb spiders, including *Argiope keyserlingi*, build silk structures known as stabilimenta around the hub of their webs. Often in the shape of a disk or a cross, the UV-reflecting decorations are especially noticeable to many arthropods and birds able to see in UV. As stabilimenta might attract predators to an otherwise inconspicuous web, this counterintuitive behavior has led to a 130-year debate over their function.

Proposed purposes include providing mechanical support for the web, giving shade, protection against predators through concealment or predator distraction, making the web conspicuous so that birds don't damage it by flying through it, and attracting visually oriented prey. While stabilimenta may provide multiple functions, evidence is leaning toward these structures being distracting to predators, while providing some improvement in prey capture.

WHY CROSS THE WEB?

A. keyserlingi uses non-capture silk from the aciniform glands to produce their often X-shaped stabilimenta. Some studies suggest that their color properties attract pollinating insects, so spiders that produce more decorations get more food. Other work suggests that stabilimenta can halve bird strike, although this is so uncommon that it seems an unlikely selection force for web-decorating behavior. Alternatively, stabilimenta might reduce predation. Bird attacks are reduced when *Argiope* is located behind a stabilimentum, possibly by

shielding it from clear view. Similarly, decorations may act as a decoy to confuse predators, such as birds, assassin bugs, and wasps, as to the location of the spider and divert attack.

WHY NOT CROSS THE WEB?

Not all *Argiope* build decorations, and some studies suggest that prey intake is lower with stabilimenta than without. Varying their size and shape may prevent potential prey from associating the decorations with danger—or, conversely, prevent associative learning by predators of a stimulus that would aid them in locating *Argiope*. Changes might simply be because the spider does not have sufficient energy or silk reserves to invest in making large decorations or decorating at all, since aciniform silk is also used in their wrap-attack method of subduing prey.

Complicating things is the fact that *Argiope* build different shapes depending on age: Often, juvenile *A. keyserlingi* build spiral or circular decorations, switching to linear or cross-like structures as adults. Most shapes to some extent shield the spider from view, so why and how there are developmental changes in the shape of stabilimenta remains a mystery that will hopefully be solved in the coming decades.

→ Courting males produce vibrations by shuddering in the female's web. This delays predatory behavior in the female and lessens the risk of precopulatory cannibalism.

Indian velvet spider

Sacrifical mother

SCIENTIFIC NAME	*Stegodyphus sarasinorum*
FAMILY	Eresidae
BODY LENGTH	⅓ in (7–8 mm)
NOTABLE ANATOMY	Longitudinal white stripes flanked with six pairs of dots
MEMORABLE FEATURE	Highly social and cooperative spider

Eresid spiders in the genus *Stegodyphus* show varying degrees of social behavior, ranging from periodically to permanently social species, with these also including solitary dispersers. Social *Stegodyphus sarasinorum* females not only regurgitate food for the young of other females in the group, thereby increasing the colony's reproductive success, but also sacrifice themselves to be eaten by the spiderlings of other mothers! Multiple generations of *Stegodyphus* are involved in upkeep of the collectively built web. Construction can take days, and maintenance is key, as the web provides a food source and protection from predators and the environment.

FAMILY MATTERS

Sociality in spiders evolved from nonsocial ancestors. In social *Stegodyphus,* especially in *S. sarasinorum,* one or more spiders attack a prey item, which is shared among the group. This cooperation increases capture success and the efficiency with which nutrients are extracted from prey, and may reduce the costs to any given individual hunter in the group. However,

to the detriment of the common good, individuals act in their own self-interest to eat the food. While this can lead to the breakdown of cooperative behavior, family ties enable cooperation by making spiders less selfish. Compared with spiders in groups of mixed relatedness, full-sibling groups of subsocial *Stegodyphus* are faster to attack prey and recruit other spiders to form feeding groups, are more efficient in extraction of nutrients from prey, and are also less aggressive. They therefore mature faster, and have more opportunities to reproduce earlier.

PREY CAPTURE AND SOCIALITY

Social species of *Stegodyphus*, unlike solitary *Stegodyphus*, are restricted to tropical and subtropical habitats that are rich in insect life throughout the year. To sustain the food requirements of social life, a minimum threshold of prey size and especially prey abundance are key factors. The importance of food on social living in clear: Compared to fed spiders, hungry social *Stegodyphus* invest more in web-building, especially of the prey-catching cribellate silk; but contrary to common assumption, larger social groups do not exhibit a reduction in silk investment. While in tiny *Anelosumus* the ability to capture larger prey is a key driver for social behavior, in *Stegodyphus* it is their ratio to that of prey that reinforces social living. *Stegodyphus* are fairly large, and social *Stegodyphus* can reduce maturation time by hunting sizable prey with a comparably smaller body size than subsocial *Stegodyphus*.

→ Many species take advantage of the sturdy structure of *Stegodyphus sarasinorum* colonies. Many foreign insect and spider species (and some vertebrates) also reside in the colonies, sometimes even eating the host spiders.

TETRAGNATHA ELONGATA

Elongate stilt spider

River spy

SCIENTIFIC NAME	Tetragnatha elongata
FAMILY	Tetragnathidae
BODY LENGTH	Females ⅓–½ in (7–13 mm), males ¼–⅓ in (6–8 mm)
NOTABLE ANATOMY	Very long and thin spiders with long narrow legs and large chelicerae
MEMORABLE FEATURE	Web spider often associated with rivers

Few spider groups provide a better indication of ecosystem health than long-jawed orb-weavers like *Tetragnatha elongata*, particularly along the edges of waterways that have tight coupling between terrestrial and aquatic ecosystems. Waterways are often contaminated with long-lived chemicals hazardous to life. *Tetragnatha* feed in these riparian habitats and are considered a monitor group, indicating the scale of pollution by their presence or absence, and a sentinel group, used to fine-scale monitor and manage contaminants in aquatic ecosystems.

RIPARIAN SPIDERS AS BIOINDICATORS

Globally distributed and favoring habitats associated with water, species in the genus *Tetragnatha* have been used as contaminant sentinels in several ways. They form part of ecological risk assessments monitoring songbirds and are used to detect the flux of aquatic contaminants, including metals such as mercury and organic chemicals like carcinogenic polychlorinated biphenyls (PCBs), to terrestrial ecosystems. *Tetragnatha* prey heavily on adult emergent insects that spend their larval period in the water and consequently retain,

or bioaccumulate, the aquatic contaminants transported by these insects. This helps determine the potential exposure of terrestrial predators to waterborne contaminants. *Tetragnatha* are also often used to monitor changes in contamination levels due to a specific disturbance, such as land use change or a point source contaminant spill. A little spider can tell us a whole lot about the health of our surroundings.

YOU ARE WHAT YOU EAT

As typically horizontal web-builders, tetragnathids often build their traps over streams or in low-hanging branches extending over water to capture prey. In turn, these spiders provide food for terrestrial organisms, such as birds. Therefore, any contaminants that accumulate within spiders are likely to be passed on to higher trophic levels, with birds then eaten by others. While in theory, spiders do not exhibit significant adverse effects of these contaminants, few studies have investigated this. If it's true that you are what you eat, it is useful to know what is inside the food being eaten. PCBs have long been banned from many countries, including the United States and the European Union because of their dangerous tendency to degrade very slowly and biomagnify, or increase at each higher trophic level—meaning that the cat that eats the bird that ate the spider that ate the mayfly could even have lethal levels of PCBs.

→   Living near water poses the risk of falling in. No matter, held by surface tension, elongate stilt spiders perform circular orienting movements to locate the nearest shore.

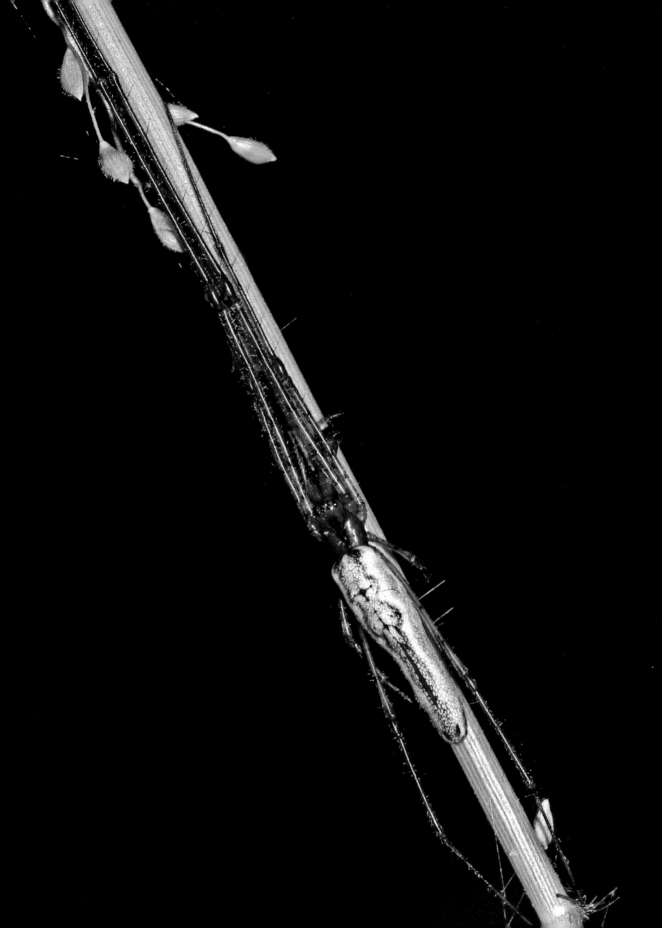

SPIDERS
AS PREY

What eats spiders?

A nursery song named "There Was an Old Lady Who Swallowed a Fly" lists the titular lady sequentially swallowing a series of animals (including spiders), to "catch" the previously swallowed one. This clearly goes beyond the biological meaning of trophic levels. Yet, the song nicely illustrates this concept and mentions commonly recognized spider predators: birds. Yet, we, too, eat spiders. In parts of Asia and South America, they are roasted, fried, or pickled. However, human consumption of spiders is minor, and spiders' defenses have evolved in response to other predators.

SPIDER PREDATORS

Soft-bodied, often with a juicy abdomen, and unversed in stinging, kicking, or other forms of retaliation—and with web spiders sitting ducks waiting to be plucked out of the web— spiders make for a rather nice meal option. Predators such as birds, bats, and lizards can almost halve spider numbers. As if that were not enough, spiders are also prey for other spiders and predatory insects, such as ants, wasps, and assassin bugs.

Spiders are also host species to numerous parasites, such as mites and parasitoids. Parasites differ from parasitoids in that they live off nutrients "stolen" from the host without killing it, while parasitoids eventually end up killing the host. Both "relationships" are known as parasitic of the "parasite" (both parasites and parasitoids) toward the host, because the "parasite" benefits at the cost of the host. Often showing considerable species specificity in selecting their host, it is primarily wasps and flies that are parasitoids of spiders. The parasitoid lays one or more eggs inside or on the spider, of which only one larva typically hatches. As it grows and develops, the larva consumes the host, essentially "sucking it dry." Finally, once the parasitoid insect is ready to pupate into an adult, it leaves the host, which is now dead, to form

→ At night, spiders have to contend with efficient night-active predators, such as bats. Bats can detect spiders based on the reflected echo of the calls they emit (echolocation) as they fly in search of food.

↙ It isn't just birds and bats that eat spiders; humans do too, as illustrated by this Cambodian dish of fried spiders.

↓ By day, insectivorous birds do not distinguish between insects and spiders—they are all juicy prey.

a pupa and develop into its final adult form. However, this is not the only way that parasitoids can use spiders as hosts. Other parasitoids specialize in attacking spider eggs, and yet others develop outside the spider's body. Whatever the lifestyle of the parasitoid, the host spider will not survive.

DEFENSIVE TACTICS

With attacks on all sides, one may wonder how there can be any spiders left at all. The answer is multifaceted. Spiders typically lay a large clutch of eggs within the protective wrapping of the silken egg sac—itself often within a silken nest. Many of these

eggs, or the spiderlings within, will not survive, but some will. The egg sac itself seems to be a strong deterrent for many predators that might try eating the eggs, but it is not as effective at deterring parasitoids, which can take a heavy toll on the eggs within. Similarly, the communal webs of spiders that aggregate may facilitate defense against predators, and this is one function that may have led to the evolution of sociality in spiders.

Individual spiders use a large array of tactics to avoid predators, the key one being camouflage. Nevertheless, it is not easy to determine spider mortality and its causes in a natural setting, and

↖ Spiders can also be the target of other arachnids: mites. Here a trapdoor spider, *Damarchus workmani*, is infested with pale mites.

↑ The eggs being carried by this female daddy longlegs pholcid spider are infested with parasites. The eggs should be pale. Instead, inside are the dark forms of developing parasitic insects, possibly wasps.

estimates may be somewhat high. The typical way to measure the impact of a given predator, such as lizards or birds, is through exclosure studies, in which an area containing spiders is netted, preventing certain animals (like lizards or birds) from accessing the area. While this enables us to determine general patterns, it is also possible that spiders, detecting that the netted area is safer, migrate into the exclosure. This will artificially increase the relative abundance of spiders within the netted area compared with the outside, and so potentially artificially inflate the detrimental effect of the predator being excluded. Nevertheless, these studies do indicate that, however you look at it, spiders are on the menu of a lot of animals.

↑ Many spiders protect their eggs not only by covering them in layers of soft silk, and then wrapping the bundle in tough protective silk, but also by carrying the egg sac around with them.

↗ Alternatively, spider egg sacs can be firmly attached to the substrate and enveloped in waterproof, exceedingly impenetrable silk, to protect the vulnerable eggs inside from predation or parasitism.

Avoiding attack

The best way to avoid becoming someone's meal is for that someone to have no clue that you are there. Spiders reduce the chances of being detected by predators in the first place using several tactics, including hiding and looking inconspicuously like the background, or crypsis. At a more basic level, they also avoid being active at a time of day when most predators are active, and so many spiders are nocturnal.

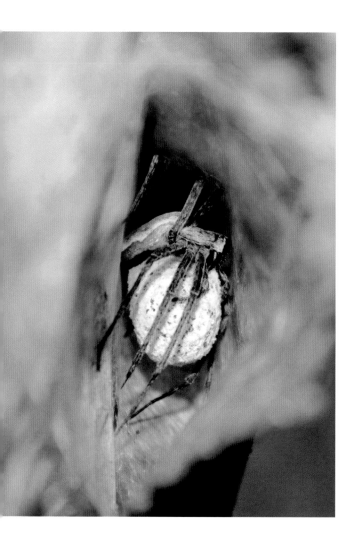

Undoubtedly the most effective way to avoid detection by a predator is to flat-out hide: for example, under bark, in a rolled-up leaf, or in a burrow. Like rabbits, many trapdoor spiders build emergency exits from their burrows, so in the case of a predator finding the burrow and entering it, the spider can still make a somewhat hasty retreat. Others have a "door" flap that, giving rise to their name, can be pulled down: It is covered in soil and debris perfectly matching the ground, making the burrows almost impossible to detect. In addition to the possible role of stabilimenta to camouflage spiders, web spiders often build a little silken retreat in their webs, typically near the branches and twigs that support the frame of the web, in which they can hide; or they can include a leaf in their web, which they roll up with silk and use as a shelter instead. These shelters are very effective at hiding the presence of spiders against predators such as lizards.

↗ *Nemesia* trapdoor spiders hide in burrows with "doors" that perfectly blend in with their surroundings.

→ Tree crevices make for good hidey holes for spiders to escape being spotted.

← Being a good mother means protecting your young, for example, inside this rolled-up leaf.

COLOR CRYPSIS

Crypsis is why so many ground-dwelling spiders
are brown, while spiders that live on leaves are often
green—but this only works against visually oriented
predators. Some of the most remarkable examples of
crypsis are found among hersiliids, aptly named tree
trunk spiders. These spiders have a near shadowless flat
body, and sit on the bark motionless, legs spread wide,
ready to ambush their prey, with markings that often
look identical to the bark or lichen on the bark. These
spiders are so well camouflaged that the main way to
find them is to sweep your hand across the bark until
you see the motion of the spider escaping. However,
crypsis in other sensory modalities is not only possible,
but quite likely, in spiders. Hiding and crypsis do have
one major drawback: movement breaks the defense,
making going about their lives doing activities such
as looking for food or mates somewhat problematic.

Other creative solutions that have evolved to avoid
predator attack include resembling an object in the
environment that is not considered as food. Here, the
spider is likely to be detected, but it will be dismissed
because it doesn't seem especially tasty. In insects, the
classic examples are stick insects, since few animals feel
the need to eat a twig. Spiders have evolved to look
like bird droppings, leaves, twigs, seeds, and dead
flowers, both in their body shape and coloration, and
also in their use of web decorations. Some species
change the color patterns on their body as they grow,
for example, by relying heavily on crypsis when they
are small and resembling inedible bird droppings
when they are larger adults.

→　To remain cryptic, a spider has to
stay motionless so as not to give away
its presence. For an ambush predator,
such as this crab spider, this is all the
more important.

CLOSE RESEMBLANCE

Many spiders avoid attack not by resembling inanimate objects in the environment, but by resembling living, yet still unappetizing objects. Being dangerous, ants are foremost in this category. Ants are well-known predators of similar-sized animals, including spiders, which tend to avoid them at all costs. It follows that looking like an ant may be a good defense against ant-averse predators. Spiders are not alone in this category. Ants can often mount a significant defense and recruit colony members to help, so many animals apart from spiders, such as birds, praying mantises, and lizards, also tend to avoid them. Hundreds of non-web-building and primarily day-active spider species have evolved a remarkable resemblance to ants: in almost all cases this has a clear protective function known as Batesian mimicry. This typically involves spiders behaving like ants, appearing to walk using six slender legs and raising the first or second pair of legs in a semblance of ant antennae; it also involves a narrowing of the body and the appearance of having three,

rather than two, body parts. Even the shiny appearance of ants is copied. This deception can be so precise for a given "model" ant species that spiders, as they grow, can change their color (for example, from jet black to red) to resemble different species of local ants that match their size at any given stage of development. These mimics do not use their appearance to deceive the ants themselves and feed on them; instead, they deceive animals that fear ants and therefore leave the spiders that look like ants alone.

← The spined crab spider, *Epicadus heterogaster*, may in some situations resemble a flower, but it often hunts on plants without flowers, and attracts prey via its ultraviolet reflectance.

→ Japanese ant mimicking spiders, *Myrmarachne japonica* (bottom), are near-perfect mimics of their ant models, *Formica japonica* (top). They even have a constriction in their cephalothorax that means they appear to have three body parts, like insects.

↓ Resembling a broken-off twig during the day, the orb-weaver *Poltys*, which builds its web and hunts by night, is a perfect example of how to avoid being eaten by resembling something inedible.

Avoiding capture

When all else fails and the predator has detected the spider, a good option is to run away to prevent being eaten; or for some spiders, dropping to the ground and remaining motionless among the leaf litter is effective. However, spiders have other solutions when things look dire. Some of these, like losing a limb or two, are rather extreme, while others, such as flinging hairs from the body into the face of the predator like a shower of tiny arrows, can be surprisingly effective.

← Spiders often feign death by rolling up and curling their legs. Known as thanatosis, this is an effective mechanism to deter predators that have already spotted the spider from actually attacking and killing it, giving it an opportunity to make a last-ditch effort at survival.

↗ For a spider at risk of capture, rapidly wheeling across the sand can prevent the predator from accurately tracking the spider's location. It is also a fast way for this *Carparachne aureoflava* spider to make an escape.

↗↗ Motion blur is an effective anti-predator mechanism: If your predator can't track you, it can't attack you. That is the basis for the rapid pumping of pholcid spiders on their webs when they are under threat.

While some spiders can run fairly quickly, few could escape predation from most of their predators, such as lizards, wasps, birds, and dragonflies, this way. Spiders that live above ground, whether on leaves, bark, or webs, will often resort to dropping to the ground if disturbed by a predator or a pesky scientist. Lying motionless in the leaf litter below for periods far extending the patience of this author, they are next to impossible to find once they have dropped, making this a highly effective way of preventing capture.

MOTION BLUR

Present in some spiders, an especially uncommon way of avoiding capture is to use extremely fast movement, either to get away from the predator or to blur the spider from accurate vision in the predator. "Motion blur" is where the visual system simply cannot keep up with the speed of motion because the neurons in the visual system cannot fire sufficiently frequently. Consider the propellors on a plane speeding up until they become a whirl in which the individual blades cannot be discerned because they are moving too fast. Many pholcid spiders, often referred to as daddy longlegs spiders (see page 190), use this when their webs are disturbed, and similar mechanisms are used by other web-building spiders, such as *Argiope* (see pages 158 and 192). The pholcids will start vibrating at such a speed that the spider essentially disappears like the props on a plane, making attack on the spider impossible. The desert-dwelling sparassid spider *Carparachne aureoflava* escapes from approaching parasitoid wasps by curling up into a ball and rolling down sand dunes at over a hundred yards per second before burying itself in the sand and becoming invisible.

← The male wolf spider courting the paler female is missing two legs. He must have escaped an attack—not unscathed, but with his life, allowing him to find this mate. Spiders are known for their ability to drop limbs (autotomy) that are in the grasp of a predator. The benefits of this extreme tactic are obvious.

↙ If you see a spider in this posture, keep away. This is a spider that is defending itself and showing you everything with which it's ready to fight.

AUTOTOMY

One of the most famous ways that spiders avoid capture by a determined predator (or indeed a vicious competitor) is by actively detaching a limb or limbs that may have been grasped by the predator, like a lizard might lose its tail. This may leave the predator with a skinny leg or two, but will allow the spider to drop or otherwise make a hobbled escape—and any option that allows it to live another day and potentially find a mate is selected. In spiders, if this "autotomy" (meaning "self-cutting") occurs before their final molt, the lost limbs can regenerate at the next molt, although the limbs will often be a little stunted. Spiders lose their legs mainly at the joints, which form natural breakage points, and they have a variety of physiological adaptations that prevent them from dying of blood (hemolymph) loss and enabling a rapid recovery at the site of the wound. Somewhat counterintuitively, the costs to spiders surviving an interaction that required them to autotomize a limb seem to be fairly minimal. Nevertheless, if the spider were to detach limbs too readily, it might end up almost limbless before its next molt—and that evidently would have detrimental effects on its survival.

DEFENSIVE BITES

Of course, if the possibility that a bite will help its survival, a spider will try to bite its attacker. Should the fangs be able to penetrate into the attacker, even if the venom is of little use, the bite itself might hurt or cause the predator to briefly release the spider, whereupon it could drop. In some spiders, such as in male Sydney funnel web spiders, venom has evolved for defense, and so may also contain compounds that are toxic or painful to the predator.

Death by a thousand cuts

Rather than succumbing to the swift kill strike of a predator, spiders may also die a very slow death through infection by another organism that causes it harm, and no discussion on spider mortality can avoid this subject. Spiders are susceptible to fungal, bacterial, and viral infections, as well as to nematode (roundworm) infections.

Numerous species of fungi are exclusively pathogens, causing disease and finally the death of spiders. These can often be detected as colorful stalks emerging from the carcass of a spider, like a miniature bonsai garden. In contrast, it is difficult to spot a spider that is suffering from a pathogenic bacterial or viral infection, yet these are common. Spider venoms have significant antimicrobial properties, which likely evolved to prevent bacterial infection entering the spider while eating. Furthermore, many spider silks also seem to have antimicrobial properties that can inhibit bacteria.

HOST–PARASITE INFECTION

Fossil evidence preserved in amber shows that the relationship between nematodes and their spider hosts dates back at least 40 million years, and this may help explain the complexity that has evolved in this host–parasite infection. Nematode infection typically displays as a spider with deformed limbs and a swollen abdomen, from which the nematode will eventually emerge—typically killing the spider in the process.

↑　A parasitic mite (also an arachnid) is an ectoparasite, feeding off its host by attaching itself to the outside of the animal. A spider may survive hosting one mite, but if attacked by several, it will not survive.

←　This spider has succumbed to a fungal infection. The external "stalks" of fungal growth emanating from the spider will be releasing spores into the environment.

A nematode-infected spider also shows behavioral changes, often becoming lethargic and seeking out water. These behavioral changes are caused by the nematode, and they facilitate the nematode to complete its life cycle, which requires it to find a source of fresh water to reproduce.

Mites, often a brilliant red, are the most common of spider ectoparasites, anchoring onto their bodies and slowly sucking on their hemolymph until they are engorged and drop off the host. In most cases, this does not cause the death of the spider, but when multiple mites attach to a single spider, it will be unlikely to survive. In contrast, parasitoids will inevitably kill their host on completion of their larval stage of development.

Spider parasitoids, of which there are more than a thousand species, can be generalists about their hosts, unfussy about which spider to attack. Alternatively, they can be exceedingly picky and only attack a certain species of spider. Like nematodes, fossil evidence shows that the relationship between parasitoids and spiders is old, dating to at least 50 million years ago. Some spider parasitoids are endoparasitoids, which develop within the host. Other species, ectoparasitoids, attach to the outside of the spider's body, out of reach of the spider, during development. Either way, they feed off the spider during this time, only killing it by feeding off of what remains of the spider when they are ready to move on to the next stage of their life: pupating into an adult. Some parasitoids specialize in attacking the egg sacs of their host, where they will likely kill off many eggs before they are developmentally ready to pupate. Parasitoid-induced mortality of spiders in infected egg sacs is severe, with up to 40 percent mortality if an egg sac is parasitized.

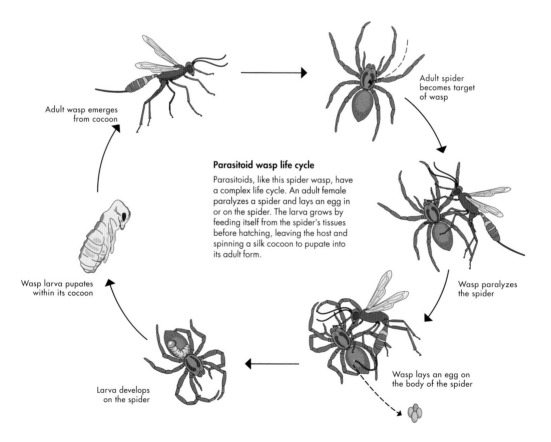

Adult wasp emerges
from cocoon

Adult spider
becomes target
of wasp

Parasitoid wasp life cycle
Parasitoids, like this spider wasp, have a complex life cycle. An adult female paralyzes a spider and lays an egg in or on the spider. The larva grows by feeding itself from the spider's tissues before hatching, leaving the host and spinning a silk cocoon to pupate into its adult form.

Wasp larva pupates
within its cocoon

Wasp paralyzes
the spider

Larva develops
on the spider

Wasp lays an egg on
the body of the spider

↑   A female parasitic spider wasp stings and paralyzes its spider host before taking it to her nest where she will lay an egg on it. The larva will feed off the paralyzed spider until it is ready to pupate.

→   Here a wolf spider has succumbed to a wasp attack.

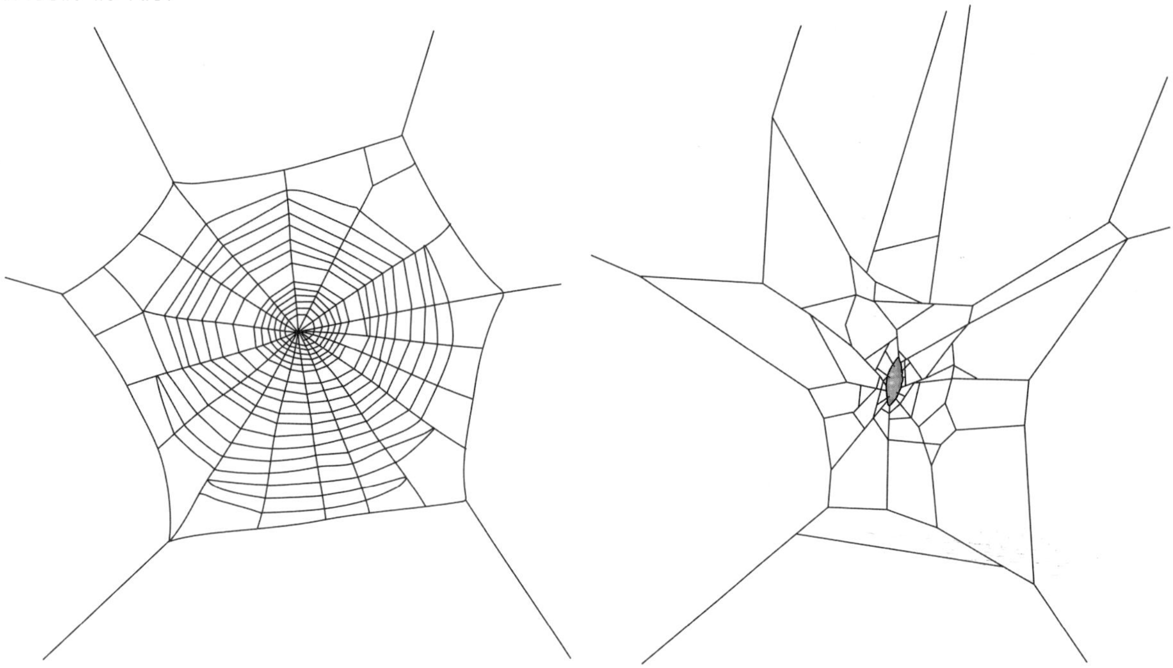

Cocoon webs
One of the most striking changes in parasitoid infected spiders is that they no longer build ordinary webs (left). Instead, they build structures that are especially strong and which serve as protection for the developing insect (right).

BEHAVIORAL CHANGES

One of the most fascinating aspects about infection, at least by parasitoids and nematodes, is that it is often accompanied by staggering behavioral changes that benefit the invasive organism at the cost of the host. Many parasites require multiple hosts during their life cycle, and for sexual reproduction to occur they need to ensure that they end up on the final, definitive, host. One way of doing this is by changing the behavior of the intermediate hosts to maximize the chances that they will encounter the final host. For example, a parasite that requires a shorebird such as an oystercatcher as a final host might manipulate the behavior of a cockle—as an intermediate host—to make it more visible because it is less able to burrow into the ground and ends up closer to the water's edge, making it more likely to be eaten by a hungry oystercatcher.

→ When parasitized, the normally social *Anelosimus eximius* become antisocial and disperse to build single webs that benefit the parasitoid.

MIND-CONTROLING WASPS

We have seen that nematode infection exerts some form of control over the behavior of the spider, making the spider more prone to seek out water. Some parasitoid wasps similarly orchestrate their host's behavior to their own survival advantage. Solitary ichneumonid wasps from the *Polysphincta* group are the stuff of web-spider nightmares (and yes, spiders might dream—see page 234). An adult *Polysphincta* will sting the spider, temporarily paralyzing it and possibly injecting it with substances that modify its behavior. The wasp then lays a single egg on the host spider. The larva, which perforates the spider to sustain itself by sucking on the spider's hemolymph, also secretes substances that affect the spiders' molting hormones, which activates web-building behaviors. The effect is that the wasp manipulates the spider such that, when the larva is about to pupate, the spider produces a modified, sturdy "cocoon web" that holds and protects the cocoon in which the wasp will develop into an adult. Having done this job, the spider dies and is eaten by the larva before it spins its cocoon.

In another example of parasitoid wasp manipulation of behavior facilitating pupation of the wasp, infected social *Anelosimus eximius* spiders suddenly get the urge to disperse from their natal colonies and build solitary and extremely dense enclosed webs. Essentially, the wasps make the spiders build them a nice safe house before they eat them to death.

Daddy longlegs spider

Leggy whirler

SCIENTIFIC NAME	*Pholcus phalangioides*
FAMILY	Pholcidae
BODY LENGTH	Females ⅓ in (7–8 mm), males ¼ in (6 mm)
NOTABLE ANATOMY	Pale, yellow-brown spiders with extremely long, spindly legs
MEMORABLE FEATURE	Likes dimly lit areas. Often prey on others spiders and exhibits a very fast whirling behavior when confronted by predators

Found in corners of the ceilings in many reader's rooms, *Pholcus phalangioides* are major predators of other spiders, but they come under attack by diverse predators themselves, including assassin bugs and even other spiders. For some spider predators, they have very effective defenses, but they seem less able to deal with the stealth of an assassin bug attack.

WHIRLING AGAINST SPIDERS

On detecting a predatory spider touching its web or even triggered by air currents, the daddy long legs spider begins to gyrate at such a speed that the spider blurs, like propellors on a plane, making attack on the spider impossible. It is quite easy to elicit this defensive response by blowing gently on the silk. However, jumping spiders alone trigger an effective alteration to this behavior by eliciting slightly slower whirling that can last several hours—yet how the nearly blind *Pholcus* discriminates hunting jumping spiders from other spiders remains elusive.

→ They might look spindly, but daddy longlegs spiders are highly effective predators of other spiders that are much heavier than themselves (such as this unfortunate wolf spider).

ASSASSIN BUGS

Using piercing and sucking mouthparts, several assassin bug species specialize in hunting spiders. Some hunt jumping spiders, while others invade the webs of web-building spiders like *P. phalangioides* without eliciting an overt response from the resident spider, sometimes by luring them toward them, mimicking the vibrations of prey caught in the web. Web-invading species must avoid alerting the spider through the very strands of silk that they must cross to obtain their prey. They do so by avoiding moving across the sticky capture area of the web. They also cut the silk lines as they approach, but this can cause an alerting ping as when a guitar string breaks, which results in attack by the resident spider.

The bugs attenuate the vibrations caused by each meticulous break by holding the loose end of the line until it sags, before releasing it and moving on to the next line—or by masking their approach by moving if there is a flutter of wind. Once within range, they reach forward and gently tap the spider with their antennae before lunging forward with the kill stab. This macabre "stroking" of the spider prior to attack is common to all spider-eating assassin bugs, regardless of whether they invade webs or ambush jumping spiders by loitering outside the doors of their nests, ready to tap and impale a spider exiting the nest. But why alert the spider? Paradoxically, this tapping seems to have a calming effect, reducing the chances of a spider fleeing from or attacking the predator, lulling it into compliance prior to its death.

Hawaiian garden spider

Master web-builder

SCIENTIFIC NAME	*Argiope appensa*
FAMILY	Araneidae
BODY LENGTH	Females 2–3 in (50–77 mm), males ⅖–⅘ in (10–20 mm)
NOTABLE ANATOMY	The large females have a strikingly bold yellow abdomen
MEMORABLE FEATURE	Webs often have stabilimenta in the form of an X

Insectivorous birds are major spider predators, clearly demonstrated by the reduction of the *Argiope appensa* population in a natural experiment. Invasive brown tree snakes have essentially driven avian life on Guam to extinction. Spiderweb surveys on Guam and nearby snake-free islands with healthy bird populations showed that spider densities were 2 to 40 times higher on Guam. While some of this may have been driven by reduced competition for shared prey based on observations and other studies, this effect was largely driven by reduced bird predation on the spiders.

LIFE HISTORY EFFECTS

Populations of all surveyed spiders, not just *A. appensa*, were dramatically higher on Guam. Other clear effects of birds on the life history of *A. appensa* were reflected in their web size, which were 50 percent bigger on Guam than on nearby islands. Similar effects of increased web size in situations of reduced predation have been found with *Argiope versicolor*, although in this case the predators were other spiders.

Since larger webs are more likely to increase prey capture, the ramifications of predation on spiders encompass more than basic survival but also alter their behavior. Not all spiders are equally susceptible to avian predation. Because many insectivorous birds are diurnal predators that rely on vision when hunting, large diurnal spiders like *A. appensa* appear to be primary target prey. Male spiders are typically smaller and more active than females, and these differences may differentially expose the sexes to predation risk, especially during months in which males are exposed because they are actively searching for mating females.

STUDYING AVIAN PREDATION

In the absence of a natural experiment like that of Guam—which has limitations because it is a sample size of one and difficult to extrapolate to a wider context—evaluating bird impact on spider communities largely relies on exclosure studies. Exclosures—in which coarse netting preventing bird entry is used to cover part of the vegetation or a large area of habitat—have often been used to compare relative abundances of spiders within and outside of the exclosure. These studies typically report at least a twofold increase in spider densities within the enclosure compared with outside, meaning that bird predation on spiders might account for roughly half of spider mortality.

→ Female *Argiope* often have striking yellow coloration, which is thought to function either to lure insects to their webs or to disrupt the body outline of the spiders, camouflaging them from their predators.

Weaver ant mimics

Danger deceiver

SCIENTIFIC NAME	*Myrmarachne assimilis*
FAMILY	Salticidae
BODY LENGTH	Females ⅓–⅖ in (8–10 mm), males ⅖–½ in (10–13 mm)
NOTABLE ANATOMY	Appearance of three body parts through constriction of cephalothorax
MEMORABLE FEATURE	Erratic walking behaviour, with one leg raised in appearance of ant antennae

Ant mimicry—like that of *Myrmarachne assimilis*—first evolved at least 50 million years ago and has evolved independently multiple times in spiders. It is currently found in over 43 genera from 13 families. In fact, hundreds of species in the jumping spider genus *Myrmarachne* alone are ant mimics.

Ants can be extremely dangerous; they often live socially and have excellent defenses, including biting, stinging, and recruiting an army to help. So many ant-averse animals, such as jumping spiders, mantises, and wasps, don't attack them. If you deceive your potential predator into believing you are an ant, when in fact you are a comparatively harmless spider, you tend not to get eaten—except, of course, by predators that specialize in eating ants.

SPIDER NURSERIES

With the appearance of having three body parts, moving frenetically on six legs with the most anterior waving and tapping, and with similar coloration, *M. assimilis* closely resemble weaver ants (*Oecophylla smaragdinda*). The spiders build their nests near ant colonies so as to capitalize on the defenses provided by these aggressive ants. This is a hazardous lifestyle, since the ants can easily overpower the relatively hapless spiders. Jumping spiders are normally a pretty antisocial bunch, but there is strength in numbers. To protect themselves and their young, *M. assimilis* often build their nests adjacent to each other and connected by silk. Experiments have revealed that eggs in solitary nests are more frequently raided by ants than eggs in connected nests, showing that these nurseries have distinct survival advantages.

OUT OF THE FRYING PAN

Compared with females, male *Myrmarachne* have massively enlarged chelicerae, which can almost double the length of the spider. Rather than detracting from their resemblance to ants, these make male *Myrmarachne* look like ants carrying something in their mandibles, so they are still safe from ant-averse predators. To ant-specialist predators, however, they resemble an ant whose dangerous biting parts are occupied. Thus fooled, the predator targets the spider because they look like an especially safe ant to attack!

Male vs. female ant mimics

Male and female *Myrmarachne* ant mimics are very similar except that males have extremely long chelicerae. This still has a function in protective resemblance, but males with larger chelicerae also win contests with rival males with smaller chelicerae, and so this trait is under sexual selection.

Female   Male

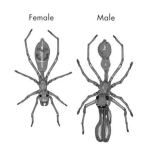

→ *Myrmarachne assimilis* spiders closely resemble ants, the most commonly mimicked animals on Earth.

Antilles pinktoe tarantula

Eighties rock star

SCIENTIFIC NAME	*Caribena versicolor*
FAMILY	Theraphosidae
BODY LENGTH	Females 1 ⅕–2 ⅖ in (30–60 mm), males 1 ⅕–2 in (30–50 mm)
NOTABLE ANATOMY	Variable features, but often bright-colored tarantula with pink "feet"
MEMORABLE FEATURE	Releases prickly hairs that cause irritation to would-be attackers

Tarantulas (Theraphosidae) live south of the southern parts of the northern hemisphere, in the Americas and Australia (New World tarantulas) and Asia and Africa (Old World tarantulas). All tarantulas rear up and extend their fangs in a threat display, but only Old World tarantulas frequently follow this display with biting. In contrast, Central and South American species rely on the use of urticating hairs over biting for defense, perhaps because their venom is less effective against mice, their typical predators, than that of Old World species. In fact, urticating hairs are not even present among Old World tarantulas.

ROCK STAR TARANTULAS

Tarantulas are nocturnal, and most are burrowing species that seldom leave their retreats. In contrast, *Caribena versicolor* builds dense hammock-like funnel webs in rainforest trees, making it a rare arboreal tarantula. This species is famous for its bright, often iridescent, green, red, pink, and purple coloration and relative nonaggression. Although it will use its legs to rub off urticating hairs when severely threatened,

this is unusual. Unfortunately, coupled with its placid disposition, its glamorous coloration makes this species a highly sought-after spider in the pet trade, which is often illegal and is severely disrupting natural populations worldwide.

URTICATING HAIRS

Named because they penetrate the skin and induce an inflammatory rash, or urticaria, in the victim, urticating hairs are found on the tarantula's abdomen (sometimes on its palps), and measure between 0.1 and 1.3 mm. They are heavily barbed along the hollow shaft. There are many subtypes within each of seven different types of urticating hairs found among New World tarantulas, which differ in location on the abdomen and in their shape, size, and barb orientation. Sitting atop a flimsy supporting stalk, these hairs easily snap near the base, facilitating their dislodgment with specialized rubbing of the abdomen with the legs during a predatory encounter. These hairs are often released from the upper side of the abdomen when a spider is threatened, but can also be embedded within silk, such as egg sacs, molting mats, or within the silken nest hammocks of arboreal species.

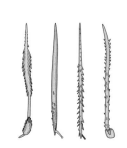

Defensive hairs
Some examples of the different morphological characteristics of urticating hairs of tarantulas, which differ in their thickness, configuration of barbs, etc. In all cases, the attachment point to the cuticle is easily breakable.

→ This Antilles pinktoe tarantula has been scraping urticating hairs from its back with its legs, leaving it with a somewhat bald patch.

Shore spider
Threat smeller

SCIENTIFIC NAME	Pardosa milvina
FAMILY	Lycosidae
BODY LENGTH	Females ⅕–⅓ in (5–7 mm), males ⅛–⅕ in (4–5 mm)
NOTABLE ANATOMY	Mid- to dark-brown cephalothorax with a distinct pale longitudinal line down the middle
MEMORABLE FEATURE	Females carry egg sacs; upon hatching, spiderlings climb onto the mother's abdomen

The wolf spider *Pardosa milvina* beautifully exemplifies the subtle way that prey can modify behavior to threat levels detected by cues from a predator. The ground-dwelling *Tigrosa helluo* is a large wolf spider that preys on *Pardosa*. *Tigrosa* relies heavily on prey movement cues to elicit predatory behavior, and in response to *Tigrosa*'s presence, *Pardosa* dramatically reduces its movement. But to do so, *Pardosa* must detect the danger. Chemotactile traces from silk or feces deposited on the ground by *Tigrosa* provides *Pardosa* with all that it needs to know.

A TASTE OF THREAT

Tigrosa females, which are larger than males, can be 30 times heavier than *Pardosa* and can eat multiple *Pardosa* at one time, making them a formidable predator to avoid. When moving along the ground, *Tigrosa* leaves behind silk and feces that contain chemical substances, which, when encountered by *Pardosa*, can provide useful information. In response to chemotactile cues left behind by *Tigrosa*, *Pardosa* decreases its activity and moves away from the ground into the vegetation, where it is less likely to encounter *Tigrosa*. The information that *Pardosa* gleans from chemotactile cues is not restricted solely to the presence or absence of its predator in a given location. *Pardosa* can also detect information about how long the predator has been in the vicinity, its sex, size, hunger level, and also whether it has recently consumed other *Pardosa*.

GRADED ANTIPREDATOR BEHAVIOR

Spending its entire time avoiding nonexistent or low-level threats would preclude *P. milvina* from going about its business of finding its own food and mates, so *Pardosa* only responds to *Tigrosa* cues when needed. For example, old cues do not elicit a response—*Pardosa* only exhibits antipredator behavior when chemotactile cues are recent, indicating that the predator may be nearby. *Pardosa* also only responds with avoidance tactics when the cues, as judged by the amount of silk and excreta detected, indicate that the "resident" *Tigrosa* is at least as large as *Pardosa*—otherwise, the prey might become the predator, with larger *Pardosa* easily attacking smaller (juvenile) *Tigrosa*. Additionally, because the large female *Tigrosa*—and those that have recently fed on *Pardosa*—are especially dangerous, *Pardosa* is strongly motivated to avoid them, using chemotactile information to determine this and detect the level of threat posed by the predator.

→ By first carrying first their egg sac and then their spiderlings, shore spiders exhibit high levels of parental care. They will even pick up abandoned egg sacs and care for the fostered young.

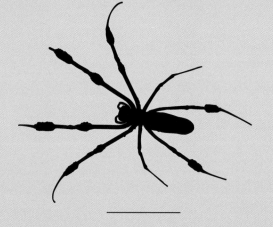

COGNITION

Working intelligence

Knowledge of spider behavior over the past three decades has completely overturned previously held assumptions that spiders are rigid, instinct-driven animals with limited cognitive capacity. Spiders are now among the best animals in which to explore topics pertaining to cognition, decision-making, learning, numerical competence, and the associated physiological and phenotypical traits, or other observable implications that these abilities may entail in animals with minute brains, even comprising sleep.

Many factors influence cognitive capacity, including individual genetic variation, social behavior, brain size, how an animal finds its food (it is easier to find grass in a grassland than to hunt gazelle in the same grassland), and the structural complexity of the environment, such as a salt plain or a tropical rain forest. The complexity of the habitat in which an animal lives affects individual cognitive capacity through developmental plasticity; spiders reared in more complex environments develop more exploratory behavior. Environmental complexity also affects cognitive capacity at a species level, providing a feedback loop between genetic and environmental factors on cognition. Relative to us,

→ As their name implies, kleptoparasitic spiders, such as this *Argyrodes* dew drop spider, parasitize the webs of other (much larger) spiders and steal the prey that is caught on the web. These skillful thieves live a dangerous existence as, if detected by the host, they could end up dead.

← A web spider busies herself repairing her web. Minor breaks of a few threads are easily fixed, and preferable to starting from scratch.

→ Using claws at the tips of each of her legs (tarsal claws) to carefully and strategically hook specific strands of silk, the spider carefully tensions her web to respond to the information that it provides her with.

spiders have unimaginably small brains, but an adult spider might have a brain ten times larger than it did as a spiderling, so comparisons within species with different brain sizes are possible. Additionally, since spiders inhabit sparse habitats that are spatially quite simple, all the way through to the most complex of habitats, the role of the spatial structure of the environment and how the spider uses it based on its sit-and-wait or active mode of predation, may be especially important in shaping spider cognition.

COGNITION DEFINITION

We might think of cognition as only encompassing learning, which is a long-term change in behavior due to experience, but this is not quite correct; therefore, it is useful to clarify what we mean by cognition. In biology, cognition is usually considered to be the mechanisms by which animals obtain, store, process,

and act upon information, which is typically gleaned from the environment through sensory receptors. This definition includes information processing at a neural level, attention, perception, decision-making, learning, and memory, but it does not require us to infer consciousness. However, this definition does lead to behavioral plasticity, in which an animal's behavior is liable to change, for example, over time in different scenarios. So, for example, male spiders may have the ability to detect (through sensory systems) differences in the size and potential fecundity of two nearby females; they can process this information and pay more attention to one of these, deciding to court the larger female. Furthermore, the male may learn some tricks to avoid being rebuffed or attacked by large females in the future, meaning that the male has stored information and learned from this initial experience. The male has therefore satisfied all criteria of cognition and then

some, since through his learned experience, he has also performed some type of risk assessment about being attacked by a female, which affects his decision-making during future interactions.

INNATE FLEXIBILITY

Another aspect of behavior evident in spiders is that instinct-driven, or innate, behavior need not be rigid, nor simple. In fact, in animals in which some aspect of the environment important to the biology of the spider is stable over many generations, it makes sense for selection to act on the spider to *not* require too much learning about that aspect of the environment. This is because learning takes up neural hardware and is metabolically costly, meaning that for an animal with limited neural capacity due to fewer neurons fitting into its small body, any shortcut that frees up neural processing will be selected for. In broad terms, selection won't act on aspects that are too changeable to be passed on, so if a spider happens to be born with an innate ability to exploit a stable aspect of its environment, it is freer to better learn about changeable aspects of its environment than its counterparts. Because it has more scope for processing with a finite number of neurons, the spider with that innate behavior is more likely to be successful and survive, and so pass on those innate traits. This is how behavior (including the ability to learn) can be genetically encoded, and it is why innate behavior is extremely useful, especially in animals that are particularly constrained by the size of their nervous system.

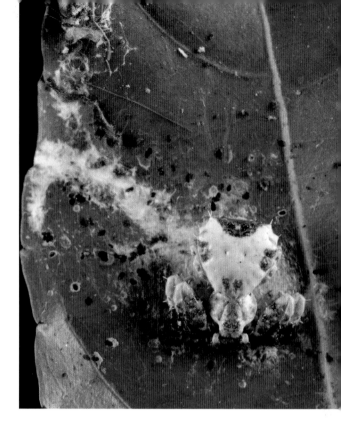

↗ Much seemingly intelligent behavior, such as alignment with objects in the environment for better camouflage, are purely instinctive and driven by sensory processes that do not involve highly cognitive behavior.

← Spiders have species-specific innate behaviors that they can adapt as needed. For example, some species will build their nests within rolled-up dead leaves suspended between foliage. Depending on the habitat, space, and leaves available, they can be flexible about which leaves to use, where to place them, and how to roll them.

However, this does not mean that learning does not occur, even with respect to innate behavior, thus promoting behavioral flexibility. Indeed, learned improvements of innate behavior—whether that is how to subdue a fly or court a female—are expected, just as a human mother with typically innate capacities for nurturing their firstborn may improve the second time around.

Despite the plasticity of innate behavior and of the potential flexibility conferred by learning, evidence is now mounting that spiders are also capable of even more sophisticated cognitive feats, including at least some basic numerical ability and the capacity to plan their actions ahead of time. One interesting aspect of spider cognition is that all the processes described above may be thought to occur within the bounds of the nervous system; but arguably, with silk, spiders may have broken this assumption, in effect extending their nervous system beyond the physical boundaries of their own bodies.

Learning and attention

Classical conditioning refers to the learned association between a strong stimulus, such as a food reward, with a neutral stimulus, such as a sound that is quickly learned to predict the arrival of food. Dog owners will be familiar with how dogs, on hearing a leash or a bag of cookies being moved, will expect a walk or dinner respectively. Similarly, in only a single experimental trial, spiders can learn to predict the imminent arrival of food after a non-related odor is presented.

These kinds of associations are very useful: Spiders can learn preferences for specific prey based on recent positive experiences, reducing their chances of encountering novel and potentially toxic or dangerous prey. On the flip side, in aversive conditioning, spiders can learn to associate a color, for example, with noxious prey. This is the basis of aposematic coloring, where the learned association between bright, contrasting colors is used to signal potent (often chemical) defenses, as found among many potential spider prey such as caterpillars and ladybugs. From a predator's point of view, a second predatory attempt at an aposematic animal would at least be unpleasant and could be deadly, so aposematic signals are a very effective way to communicate a potentially lethal threat.

← Animals that contain toxins — like this blister beetle, with its irritant and blister-forming toxin cantharidin — advertise their dangerous attributes using bright, contrasting warning (or aposematic) coloration that effectively says: "do not touch!"

↗ Some jumping spiders, such as this *Portia*, invade the webs of other spiders to attack the resident. They do so carefully, but also cleverly: They pluck the web to lull the resident into a false sense of security.

EXPERIENCE, THE GREAT TEACHER

Previous experience shapes spider behavior in
a multitude of ways. Experiments have shown that
male combatants with previous experience of losing
battles against other, larger, males, typically lose
subsequent fights even against spiders of the same size,
while females may choose to mate with males of
a familiar coloration pattern to the ones they have
previously experienced. Spiders may also use trial-and-
error learning to determine the best tactic for a given
outcome. For example, jumping spiders in the genus

Portia (see pages 60 and 232) have the dangerous
dietary preference of eating other spiders, many of
which are web-builders. To do so, *Portia* will often
invade the web of the resident spider and gently
strum on the silk to lure the spider slowly toward the
movement—sufficiently close for *Portia* to make a kill
lunge. Some types of plucking may trigger the resident
to approach dangerously fast and aggressively, which
could end up with the strummer becoming the meal.
Portia use trial and error to derive the most effective
strumming pattern of the web for the given type of

REVERSAL LEARNING

One trial has shown that *Marpissa muscosa* (see page 228) can learn to associate either color or location with a food reward and overwrite previous associations (reversal learning). Here, a drop of sugar water is hidden behind one of two identical L-shaped Lego arrangements, one blue and one yellow. Once the spider learns where the reward is, the location of the yellow and blue Lego is swapped, and the reward is either hidden behind the same color or in the same location with a color swap. In other instances, the reward is hidden behind Lego in which both location and color have changed. Spiders learn these complex associations after only a single trial.

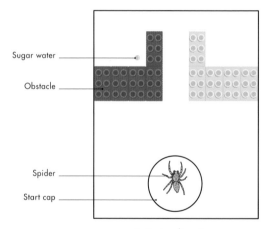

Sugar water
Obstacle
Spider
Start cap

1. Start configuration

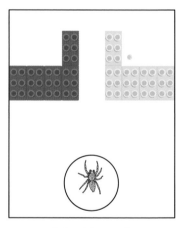

2. Complete reversal

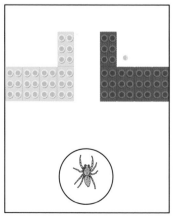

3: Color reversal

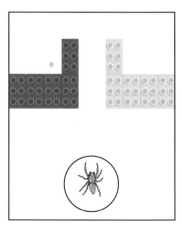

4: Location reversal

prey spider. Similarly, the trashline orb web spider *Cyclosa octotuberculata* (see page 230) uses prior experience of where its web most frequently intercepts prey and positions itself, applying tension to the silk in that area with its legs, the more easily to respond quickly to the prey it wants to catch. *Cyclosa octotuberculata* is, in fact, learning about locations in space and exhibiting spatial cognition, as well as selective attention.

SELECTIVE FOCUS

At any given time, a spider may be bombarded with potential sources of information, most of which will not be biologically relevant to the animal. Consider *Cyclosa* in its web. It is a gusty day, with variable wind from many directions shaking the web. In this vibratory chaos, some small insects may fly into the web, which should be attended to, in addition to small leaves blown into the web, which should be ignored. The process whereby animals selectively attend to relevant information, while ignoring irrelevant information, is selective attention, and this can be primed by experience. *Portia* is more likely to detect a prey spider having previously had an experience with that prey type—so experience has primed *Portia* to selectively attend to specific cues associated with that type of prey.

← Trashline spiders in the genus *Cyclosa* (see page 230) may use the "trash" decorations on their webs for camouflage or to distract predators, but they also use prior experience to build webs that are the most effective at catching prey in a given location.

Spatial cognition

Wandering spiders may learn to associate some part of the local environment, such as a distinctive rock, in the vicinity of the retreat. When returning to the retreat, it may look for that marker, just as we use landmarks to help our navigation. These cues are often coupled with other mechanisms, such as draglines left by the spider, Hansel and Gretel-like, on its outbound journey, or more sophisticated navigational methods, such as dead reckoning or path integration, where an animal estimates its position by monitoring direction and distance.

INTERNAL GPS

Essentially, path integration is an internal representation of the position of the individual with respect to a fixed origin, although it is possible that a similar mechanism might be used by some spiders with respect to a goal rather than a fixed origin. This internal representation frequently updates the distance and direction traveled, meaning that the animal can integrate this information to obtain a vector to a known point without the need for high-level conscious thought— or the ability to do calculus! Thus, after a circuitous outbound route, spiders can return to their retreat or burrow in a straight line by using path integration,

← Wandering spiders may use prominent features in their environment as beacons or landmarks to relocate their burrows after a night on the prowl.

→ Having caught prey, actively hunting spiders will often return to their hidden nest or burrow to consume their food in safety.

as found in some burrowing spiders such as wolf spiders, the ctenid spider *Cupiennius salei* (see page 56), and the funnel web spider *Agelena labyrinthica*, among others.

SPATIAL PROWESS

Advanced spatial ability is also found among jumping spiders. This is illustrated by their use of indirect detours to reach a goal for which a direct approach is unavailable. Detours can consist of several subsections: A jumping spider walks up a tree trunk connected to a branch, which connects to a twig, which connects to a leaf. A blowfly (the goal) sits on the leaf, but the spider can't fly directly to the leaf. To reach the blowfly, the spider must take a circuitous route to get to the leaf and is likely to lose sight of the blowfly in the process of getting there. This suggests that it must retain some form of internal representation (or memory) of the blowfly and its location relative to the spider for the duration of the detour, possibly through some mechanism akin to path integration. Furthermore, jumping spiders are quite capable of using detours that—counterintuitively—may initially lead them farther away

Path intergration

An animal that uses path integration is able to wander about its environment and keep track of its compass direction and the distance it has traveled. With that information, it can calculate a "home vector," or a direct path to its point of origin. Here, the spider leaves its burrow and wanders until it encounters a beetle. Upon returning to its burrow, it is able to work out a direct route using the information it gathered on the way.

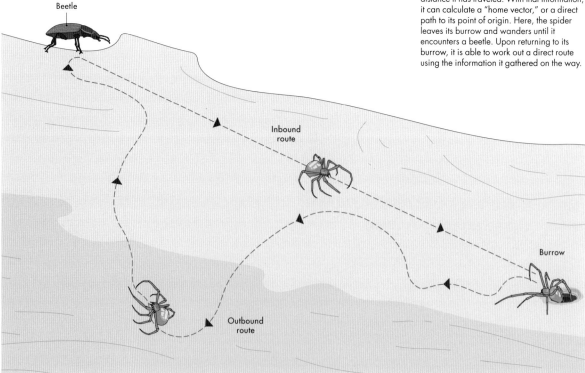

Beetle

Inbound route

Outbound route

Burrow

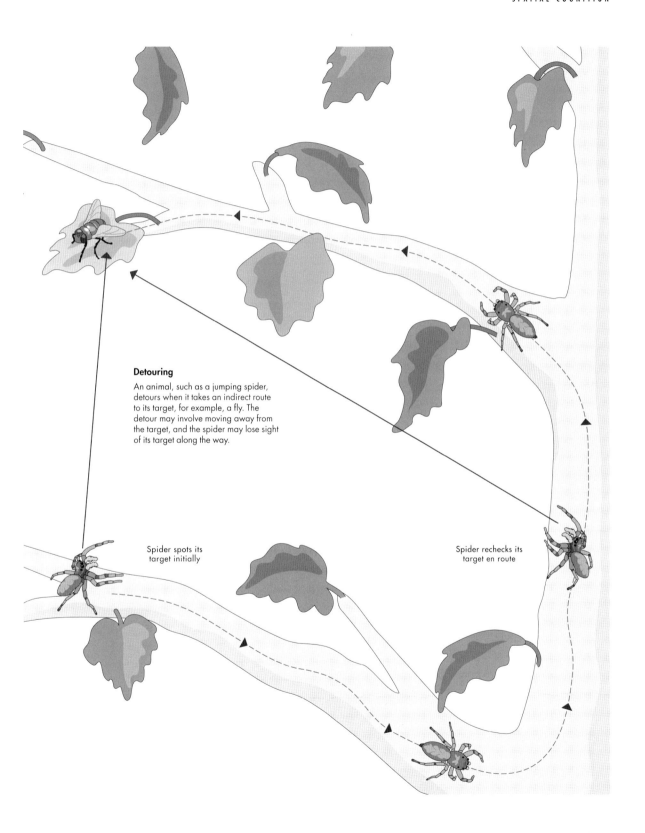

Detouring

An animal, such as a jumping spider, detours when it takes an indirect route to its target, for example, a fly. The detour may involve moving away from the target, and the spider may lose sight of its target along the way.

Spider spots its target initially

Spider rechecks its target en route

from their goal, simply because that is the only option that will eventually lead to the desired target. This suggests that there may be multiple routes that could potentially lead to the blowfly, or there may be some routes that lead near to the blowfly but not quite, so the spider may be assessing alternative options and choosing between them. All these conditions have been demonstrated in detour-taking jumping spiders.

↑ As their name implies, jumping spiders jump to attack their prey. Sometimes the best angle of attack requires them to approach from a certain direction. They may use detours to approach their prey, sometimes even losing sight of the target during the detour.

HOMING

Homing differs from path integration in that is it somewhat simpler. Where path integration requires a mechanism akin to the spider having a combination of a map and a compass, allowing the animal to navigate to a location regardless of its starting point, homing is like having a basic map with no compass. Imagine that a spider, having left its burrow and wandered about, is caught by a scientist and displaced to the side by three feet: A homing spider will travel "back" in a straight line to where its nest would have been had it not been displaced, so it will be three feet out on its return. It lacks the ability to determine a home vector and change its compass angle to get back. This hypothetical example is a common way of testing whether animals use homing ability or the more sophisticated path integration ability. Few spiders have been tested for homing ability, but this behavior is known among the nocturnal spider *Leucorchestris arenicola* (see page 224), and the wolf spider *Lycosa tarantula* (see page 226).

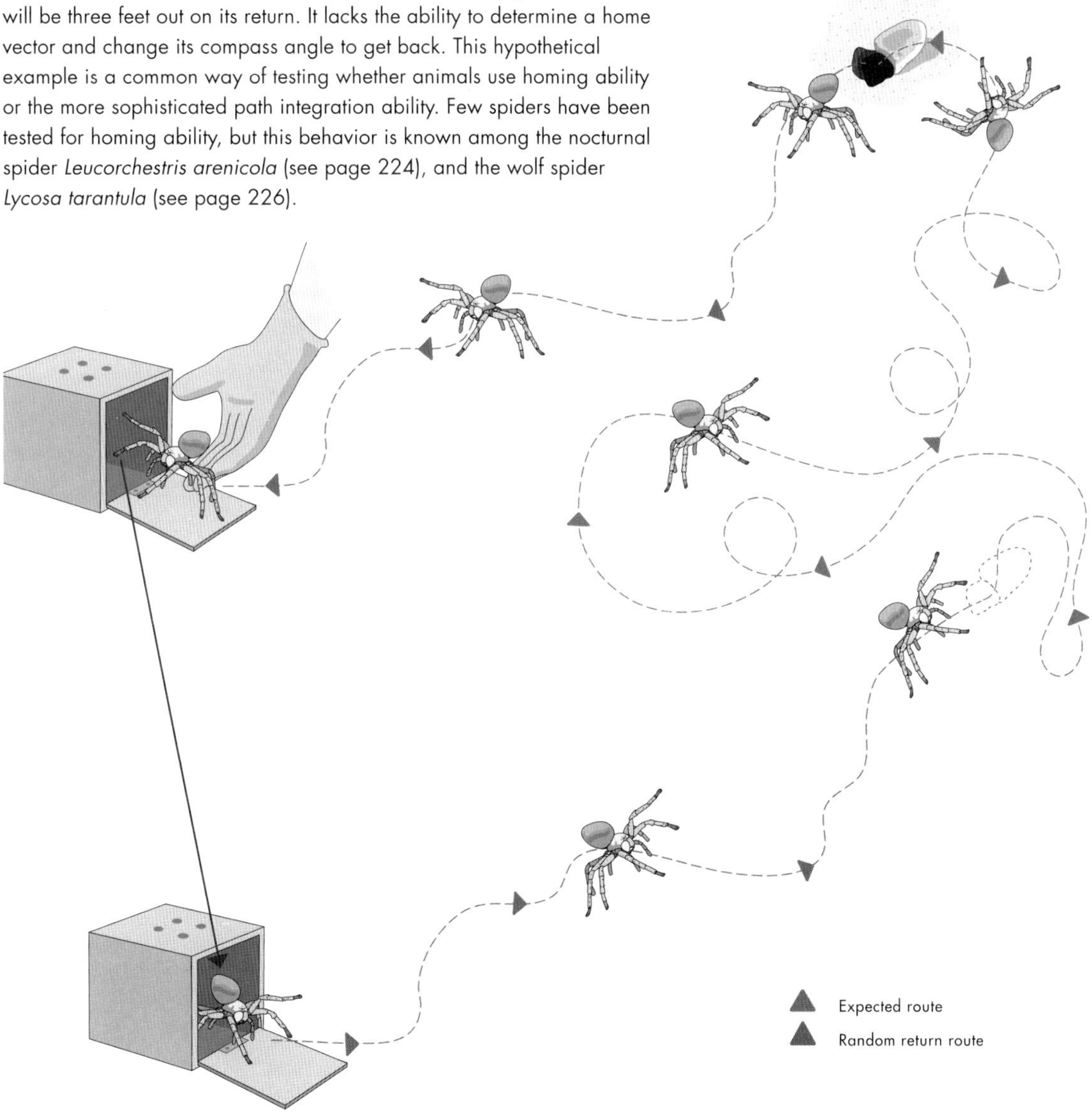

▲ Expected route

▲ Random return route

Decision-making, planning, and the gnarly concept of numbers

The spatial abilities of spiders do not end with how to return to a retreat or learn from experience where a web is most likely to intercept prey, but also extend to planning ahead, even in the absence of prior experience. Experiments suggest that some spiders, with brains smaller than ⅖ in (1 mm), can assess and plan routes ahead of time.

↘ *Portia* being tested for its spatial ability and their prowess at forward planning in a water maze.

Portia fimbriata (see page 60) and *Trite planiceps* jumping spiders were placed on a small platform in an arena surrounded by water (to which they are averse) with interspersed dowels used as a variety of escape routes that differed in length or distance between dowels, simulating safer or riskier escape options. The idea is that when inter-dowel distance is the same, the shortest route shouldbe taken, since it reduces the chances of the spider falling into the water should it fail to land a jump between dowels. In contrast, if one route has shorter, almost walkable, inter-dowel distances compared to the others, then one is the least risky and should be chosen. The catch was that spiders had to make their visual assessment of all routes before leaving their central platform, because in all cases the first jump was a big one, and so they were tested for their ability to account for risk in forward planning. The spiders were generally pretty good at this problem, but species-specific differences in spatial ability were found, with *Portia fimbriata*—which in nature lives in more spatially complex habitats than *Trite planiceps*—taking much longer to assess routes prior to leaving the central platform and correspondingly making better decisions than *Trite*. In these experiments, experience played no part: The spiders had never encountered a situation like this before, yet they appear able to plan ahead and make decisions based on the riskiness of each route.

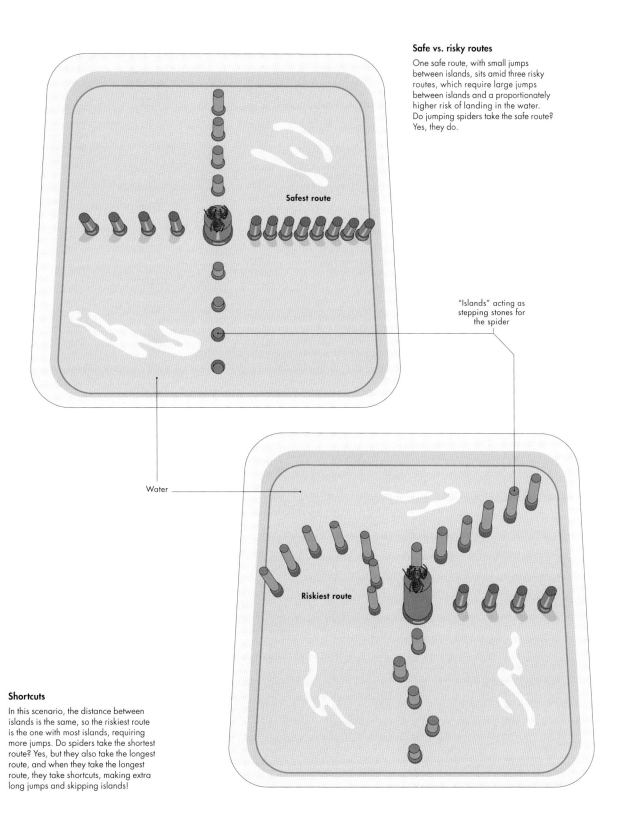

Safe vs. risky routes

One safe route, with small jumps between islands, sits amid three risky routes, which require large jumps between islands and a proportionately higher risk of landing in the water. Do jumping spiders take the safe route? Yes, they do.

Safest route

"Islands" acting as stepping stones for the spider

Water

Riskiest route

Shortcuts

In this scenario, the distance between islands is the same, so the riskiest route is the one with most islands, requiring more jumps. Do spiders take the shortest route? Yes, but they also take the longest route, and when they take the longest route, they take shortcuts, making extra long jumps and skipping islands!

← There is evidence that spiders can keep track of stored prey wrapped up for future consumption, but even they can't control for food theft by the small kleptoparasitic (meaning parasitism by theft) spiders residing on their web!

↓ Spiders like this bowl and doily spider may wrap several prey and stash them in their webs for future consumption, but, like any hoarder, they must keep track of them for fear of being robbed.

PLANNING APPLICATION

The notion that a spider may be able to plan a strategy to reach a goal ahead of time may seem far-fetched, but there is now strong evidence—at least among some species of the highly visual jumping spiders—that this is the case. So, what about memory? Having caught an abundance of prey, golden orb web spiders, *Trichonephila clavipes* (see page 266), store excess prey in larders near the hub of the web. Each prey item is attached by a dragline, but this makes them targets for theft. Some spiders, most notably among the theridiid *Argyrodes* dew drop spiders, make their living from invading the webs of other spiders and stealing the prey stashed within. *T. clavipes* is susceptible to *Argyrodes*' penchant for stealing and so seem to monitor their larder. If prey have been removed, *Trichonephila* search for longer if the stolen food was especially large or if many prey have been taken. This suggests that they can roughly track and remember the amount, and possibly the number, of prey stashed in their larder. Similar results have also been found with other spiders, such as the bowl and doily linyphiid spider, *Frontinella communis*, and western black widow theridiid spiders, *Latrodectus hesperus* (see page 54).

To date, it has been difficult to ascertain memory retention time, with some evidence suggesting that this is shorter in juvenile spiders due to their accompanying smaller brain, than among adults of the same species. Many studies have only tested memory retention in terms of seconds or minutes, while in jumping spiders memory of up to two weeks is achievable with reinforcement. As in other animals, memory formation is likely complex and highly dependent on motivation, among other things: A relatively hungry spider may be more likely to form and retain a memory of the food in its larder than one that has eaten its fill.

QUANTITY OVER QUALITY?

The examples above allude to the notion that spiders are also capable of making discriminations based on quantitative abilities. We tend to think that the use of quantitative information is a uniquely human trait associated with language and the use of numbers. However, basic quantitative ability to make estimations without symbolic representation may be very important, for example, in comparing the amount of food available in different foraging locations. Recent work suggests that this ability extends well beyond vertebrates to include invertebrates. While quantitative ability may primarily pertain to an estimate of quantity being smaller or larger than another quantity, some studies on the jumping spider *Portia africana* (see page 232) suggest that some numerical ability, such as a more abstract difference between 1, 2, and 3, independent of the quantity, may also exist. For example, one fly beside two flies might be a no-brainer for a hungry spider: Go for more! That assumes the flies are the same size and that in the "more" option, the volume of fly is twice that of a single fly. Testing the abstract concept of number is therefore precarious, because unlike symbols, the larger the numerical amount of something, typically the more there is (such as volume, or length of a rope, or song duration, etc.). Therefore, studies of numerical ability must disentangle the number concept from the associated continuous variable with which numerosity will be associated. In the absence of training spiders to use symbols to represent numbers, other methods must be used, and they are proving enlightening.

← Spiders seem to be able to keep track of the prey in their larders, but precisely what they are tracking is unclear. Perhaps rather than a specific number of prey, they are simply aware of the combined weight of their larder on the web.

Dancing white lady spider

Night voyager

SCIENTIFIC NAME	*Leucorchestris arenicola*
FAMILY	Sparassidae
BODY LENGTH	1 in (up to 25 mm)
NOTABLE ANATOMY	Very large pale spider with long legs; can reach a leg span of 5 ½ in (140 mm)
MEMORABLE FEATURE	Nocturnal, sand-dwelling spider famous for navigating long distances to home burrow at night

This nocturnal huntsman lives in the sands of the Namib Desert, and females and juveniles rarely leave their burrows. In contrast, and only on the darkest moonless nights, adult males undertake long, meandering trips away from their burrows in search of mates. They then return in a straight line to their home. These trips can be over 4,000 times the body length of the spider, and the beeline return can be 100 yards (90 m) plus. So how do they navigate back so accurately?

HOW TO GET HOME

Homing ability and spatial cognition is well understood in hymenopterans, such as ants and bees, which make return excursions from their nest or hive. On initial nest departure, these insects often take learning walks or flights. They will make zigzag movements at increasing distances from the nest to form visual snapshots of the surrounding area. These help them fix their position relative to the nest upon return, since views on the return trip are matched and aligned with the learned snapshots. In addition, they use celestial cues from the movement of the sun across the sky and sun polarization.

Conversely, little is known about spider navigation. Because of the spiders' nocturnal activity patterns, major compass cues used by hymenopterans, such as those provided by the sun, are unavailable to *Leucorchestris arenicola*. The spiders use neither moonlight, vibrational cues, gravitational cues, or olfactory cues to locate their burrows, although star-based celestial cues remain a possibility.

COLLECTING PHOTONS

L. arenicola's ability to home is driven by learning walks performed by spiders when they leave their burrows. Presumably this helps them commit the horizon landmarks seen from their burrows to memory, as in hymenopterans, but unlike for those other species, there is almost no light available to *Leucorchestris* to make its homing snapshot, so nocturnal vision is crucial to detect coarse landmarks along the horizon. As with *Deinopis* (see page 62), *Leucorchestris'* remarkable sensitivity to light comes from large photoreceptors that seamlessly pool or merge the few photons available at night (spatial summation). However, this is not enough. When walking, *Leucorchestris* often pauses for periods of time, and when coupled with spatial summation, during these stances the photoreceptors collect enough photons over time to trigger a response (temporal summation) in the visual system. While this type of vision does not allow them to detect motion, landmarks would be visible, and this coarse-grained horizon skyline is likely to be sufficient to get the spiders to within their territory and thus their homing snapshot.

→ It is thought that fog condensing on the trapdoor web provides these desert spiders with drinking water.

True tarantula

Dance muse

SCIENTIFIC NAME	*Lycosa tarantula*
FAMILY	Lycosidae
BODY LENGTH	Females 1–1⅖ in (25–30mm), males ⅘–1 in (20–25 mm)
NOTABLE ANATOMY	Very large wolf spider with striped legs
MEMORABLE FEATURE	Burrowing spider with high levels of sexual cannibalism

Unlike *Leucorchestris* (see page 224), most spiders that roam away from their home only cover short distances. The way in which they navigate has been well studied in the wolf spider *Lycosa tarantula*, after which the Italian folk dances, tarantella, are named. These frenetic dances were believed to be a cure for the spider bite, but not only is the bite of *L. tarantula* not toxic to humans, theraphosids—another group altogether—ended up getting the generic name "tarantula."

Navigating back after leaving its burrow to find food or mates, *L. tarantula* uses path integration, rather than landmark or chemotactile information, to return. Adopting path integration, the spider keeps track of directions (including angles turned) using some kind of compass, as well as distance traveled using some form of odometer, to calculate a home vector for a beeline return.

POLARIZED COMPASS

Being largely nocturnal, a sun compass is of no great use to *L. tarantula*. To animals capable of detecting polarized light such as *L. tarantula*, the pattern of polarization in the sky can be used to estimate the sun's position—even when it is overcast or twilight, and in some animals it is visible even at night. The anterior median eyes of *L. tarantula* are responsible for polarization vision; thus these eyes are key for a global reference system to update their angular position and calibrate directionality on outbound journeys. Additionally,

L. tarantula updates its home vector by using visual information provided by the downward-directed anterior lateral eyes, which keep track of the substrate and any rotation that the spider makes with respect to the ground, thus estimating the angles required to return to the burrow.

OPTIC FLOW

Spiders can measure the distance walked in two ways: one akin to step counting and the other based on vision—optic flow. To date, there is no evidence for the former among *L. tarantula*, but like some ants, they do use optic flow. Much as we can sense speed by the number of telephone poles zipping past our visual periphery as we drive, some animals, such as *L. tarantula*, can use self-induced optic flow to calculate the distance traveled because of the motion-based visual information that is generated as they walk. In this case, yet another pair of eyes—the lateral-facing posterior lateral eyes—seems to be primarily responsible for this role, probably with input from other secondary eyes that are directed at different parts of the visual field.

→ In these relatively long-lived spiders, females can reproduce over two years, but they show reproductive senescence, whereby older females produce smaller egg sacs with fewer spiderlings than their younger counterparts.

MARPISSA MUSCOSA

Fencepost jumping spider

Lifelong learner

SCIENTIFIC NAME	*Marpissa muscosa*
FAMILY	Salticidae
BODY LENGTH	Females ⅓–⅖ in (7–14 mm), males ¼–⅓ in (6–8 mm)
NOTABLE ANATOMY	Mottled brownish jumping spider with pale band under the forward-facing eyes
MEMORABLE FEATURE	Hides under bark or stones

Learning is neuronally and metabolically costly; consequently, it is typically associated with animals with large brains. Nevertheless, many small-brained animals, including spiders, do learn. For example, after one trial, *Marpissa muscosa* jumping spiders learn to associate either a food reward or an aversive stimulus with a color (see page 210). However, there is considerable individual variation in learning ability. This may be due to personality differences: Irrespective of parentage, *M. muscosa* reared in physically or especially socially enriched environments are more exploratory and less aggressive than those reared in impoverished environments. Alternatively, learning ability may arise from the social or physical attributes of the spiders' early development.

SOCIALIZED SPIDERLINGS, SMARTER UPBRINGING

To test for developmental effects on learning ability, several broods of *M. muscosa*, each with the same parents, were split and reared in one of three groups before being given learning tasks as adults. Two groups were reared either in a physically or in a socially enriched environment, so genetic effects were minimized while controlling for the effect of rearing in isolation in physically complex environments or rearing with other spiderlings and minimal physical enrichment. The third group had neither physical enrichment nor visual contact with other spiders. Irrespective of genetic factors, early social (and to a lesser extent physical) enrichment improved associative learning ability in adult *M. muscosa*.

This correlates with findings demonstrating plasticity in the brain. In contrast to *M. muscosa* reared in solitary environments, early development within physically or socially enriched environments leads to larger overall brain volume, as well as enlargement of the arcuate body in the brain. This is thought to be one of two main areas associated with high-level integration of information, and likely to play a key role in cognitive attributes like decision-making and learning.

→ These spiders display consistent behavioral differences between individuals (personality) in activity and boldness-related traits.

Trashline orb-weaving spider

Web thinker

SCIENTIFIC NAME	*Cyclosa octotuberculata*
FAMILY	Araneidae
BODY LENGTH	Females ½ in (12 mm), males ⅓–⅖ in (7–9 mm)
NOTABLE ANATOMY	Cryptic brown spiders with a lumpy appearance that hang facing down on vertical orb webs
MEMORABLE FEATURE	Sometimes decorate webs with detritus and prey remains in linear fashion emanating from the hub

Fragile, yet crucial to obtaining food, a spider's web is paramount to its survival and reflects the owner's ability to do so. As exemplified especially well by *Cyclosa octotuberculata*, several strands of evidence show that web spiders use past experience to predict future outcomes, and the web provides a testimony of this learned experience.

WEB-BASED LEARNING

Rather than envisioning cognition purely as a central nervous system process that requires a large brain to cope with complex environments, we can see spiderwebs as extending cognition toward the environment, outsourcing information processing and reducing the need for a large brain, while maintaining the behavioral richness required of a predator.

Spiders fine-tune the geometry and tension of their webs to increase capture success tailored to specific locally abundant prey and areas of the web that intercept the most prey. *C. octotuberculata* pulls radial threads, which most effectively transmit vibrations from prey, more strongly in the direction from which it has most often experienced prey capture. Thus, previous experience of prey capture in certain areas of the web directs the behavior of the spider in a flexible manner that accords with predictions of future prey capture. This is an example of spatial learning, where instead of the learning being directed by movement of the animal, it is space within the web that is learned about, suggesting that the web is, in itself, an extension of the spider.

EXTENDED COGNITION

Small brains may lead to lower cognitive capacity and errors in processing information. Yet spiders do not seem to be cognitively limited. Take the web: The silk acts as more than an extension of the sensory system. It is an extended perceptual system, enhancing sensitivity through amplification or attenuation of specific vibrations through the active role taken by the spider. Extended cognition considers that the web's structural connections and organization are integral components of the cognitive system itself, reducing the need for centralized cognitive processing. For example, the web's radial threads modulate prey vibrations to optimize signal transmission through the web, but the threads are not merely passive sensory transmitters. Changes in radial thread tension affect signal transmissibility at different vibration frequencies. The spider alters thread tension, directly controlling its perceptual world, such that it selectively "tunes" the overall attentional system extended through the web to maximize sensitivity to distinct kinds of stimuli. Through active thread adjustments, spiders can thus process the same information in different ways.

→ The web decorations of *Cyclosa octotuberculata* include carefully placed prey remains and other debris extending linearly from the hub, or "trashline." These may function to reduce predation, perhaps through disruption of the spider's outline.

PORTIA AFRICANA

Dandy jumping spider

Predator by numbers

SCIENTIFIC NAME	*Portia africana*
FAMILY	Salticidae
BODY LENGTH	Females ⅕–⅖ in (5–10 mm), males ⅕–⅓ in (5–7 mm)
NOTABLE ANATOMY	Elongated hairs give a spiny appearance; males have strikingly broad, bristly palps
MEMORABLE FEATURE	Juveniles often aggregate in groups outside the retreat of spider prey to ambush it

A toy is placed under one of two bowls in front of a toddler. The bowls are then moved around, and when they are lifted the toy is under the wrong bowl. The toddler stares uncomprehendingly. This type of expectancy violation experiment is often used to test working memory and understanding in children. Similar methods can be used to test numerical competency—in jumping spiders.

Here, a spider is allowed to observe a scene containing a discrete number of items. This scene is then blocked from view and later comes into view again, often with a change in the number of items. How long the spider stares at the new scene provides an indication of the spider's working memory expectation of the number of items, based on the original scene.

PROCESSING LIMITATIONS

In expectancy violation tests, *Portia africana* were briefly presented with a scene in which there were varying numbers (1–4 and 6) of "lures," or items made from dead individuals of one of their most common prey (*Argyrodes* spiders) in lifelike postures. After this, *Portia* was allowed to see the scene again, and when there was a mismatch in the number of lures presented in the second viewing from that expected from the first viewing, the spiders looked at the scene for longer— but only when one of the scenes contained one or two lures.

Any scene containing three or more lures was simply too many, or at least not especially biologically relevant, like our concept of lots rather than a discrete number.

RELEVANCE OF NUMBERS

The biological underpinning of *P. africana*'s numerical ability becomes clear from experiments designed to imitate prey-specific predatory encounters by *P. africana* juveniles (adults do not use this tactic) with another prey spider, *Oecobius amboseli*. *O. amboseli* build tiny star-shaped webs on stones and reside between the stone and the web. On seeing an *O. amboseli* web, *P. africana* will approach and patiently wait beside it until *Oecobius* enters or exits, when *P. africana* will pounce. Alternatively, *P. africana* will pluck the silk and drive the resident *Oecobius* out from under the web. Among the many factors in *P. africana*'s decision to settle near a web is the presence of other *P. africana*, because if one spider captures *Oecobius*, other *P. africana* may share in the bounty. Tests show that *P. africana* prefer to settle by a web when only a single other *P. africana* is present, and they can distinguish between one, two, and too many with which to share nicely.

→ A hunting *Portia africana* uses its appendages to signal on the web of a prey spider to lure it out, determining the signal that elicits a slow, non-dangerous approach though trial-and-error learning.

Metallic jumping spider

Dangling sleeper

SCIENTIFIC NAME	*Evarcha arcuata*
FAMILY	Salticidae
BODY LENGTH	Females ⅓ in (7–8 mm), males ⅕–¼ in (5–6 mm)
NOTABLE ANATOMY	Often have pale transverse lines across the "face"
MEMORABLE FEATURE	Sometimes found hanging from a thread to sleep at night

Sleep is a behavioral state of low awareness and activity. On the face of it, this is not adaptive, and its function to this day remains elusive. It is thought that sleep is important to restore neurons, consolidate memories, and keep metabolism in check. Similarly, the function of dreams is elusive. Hypotheses suggest that dreams help to consolidate thought processes from recent experiences. In humans, rapid eye movement (REM) sleep occurs in bouts throughout the lengthier slow-wave sleeping period, and it is in REM sleep that muscle twitches and dreams prevail, so the eye movements may directly reflect the dream's content.

Activity of neurons associated with movement and direction in mice suggests that rapid eye movements correspond to changes in heading during sleep, such that REMs could indicate the visual scene experienced while dreaming. However, we only have evidence from humans reporting on dreams; for other animals, such as the dog chasing the rabbit in its sleep, we can only speculate. Yet jumping spiders seem to have REM sleep. Could they be dreaming?

→   In a manner similar to that of geckos, the legs of metallic jumping spiders have tufts of adhesive hairs, each bearing miniature filaments, providing an estimated 624,000 contact points with which they can attach to surfaces.

HANGING BY A THREAD

Unlike most jumping spiders that safely put themselves "to bed" at night within their silken retreats, sometimes *Evarcha arcuata* suspend themselves on a silk thread attached to vegetation and hang upside down—seemingly very vulnerable—at night. Unlike many animals with fixed retinae, awake and active jumping spiders direct their gaze by moving the retinae of the anterior median eyes, much like we do. As in humans, sleeping *E. arcuata* exhibit bouts of distinct retinal movements, as well as leg twitching and curling. This tantalizing finding may suggest that the spider's retinal and leg movements reflect a REM sleeplike state. However, direct measurements of brain activity have not been made, so how these might correspond to the brain activity patterns found in other animals is still unknown.

EMBRACING DIVERSITY

Sleeplike states and changes in brain activity patterns have now been found in many taxa from cephalopods through insects, to reptiles, birds and mammals. However, there is significant variation between groups, and often the observed patterns do not easily fit within the human template of REM and slow wave (or non-REM) sleep. Rather than trying to fit a square peg into a round hole, perhaps understanding this diversity will reveal deeper insight into the function and evolutionary origin of sleep. Jumping spiders look like a very promising group in which to consider these questions.

DANCING,
MATING
& DYING

Sex and the spider

Sexual behavior and reproduction in spiders is varied and complex, and their sexual behavior can only be categorized as extreme, weird, and wonderful, with more than a hint of danger. We could consider the process as beginning with males courting females, enticing them to mate with them. Although mate selection in spiders is almost always dominated by the female, which invests considerably into the care of eggs and young, male mate choice and mutual courtship behavior is also present.

COURTSHIP PRACTICES

Courtship takes many forms, depending on whether it is on a web or outside of a web, and it can involve communication in many sensory modalities. In some instances, courtship behavior is as complex as any found among any animal, including the well-known intricate dances and songs of paradise birds. Should courtship proceed to the near-mating stage, females (and occasionally males) may cannibalize the male prior to mating, or later—during or after mating. In some cases, such as in widow spiders (*Latrodectus*), the male deliberately flips over to offer himself up for consumption by the female. Of course, the female

→   Courting male spiders must approach the typically larger females with care. Mistakes could be deadly, and good communication, as in any good pairing, is vital.

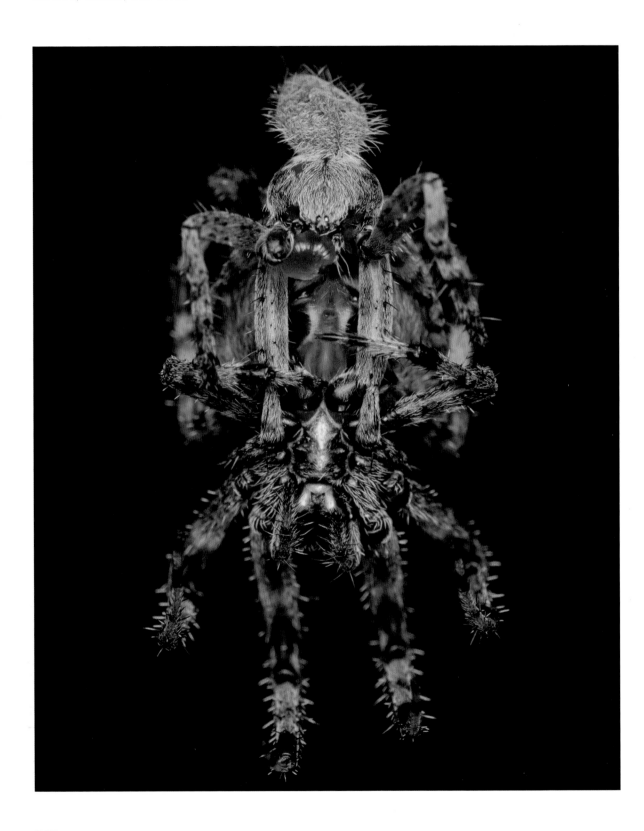

may not cannibalize the male at all, and some species exhibit much stronger sexual cannibalistic tendencies than others.

Similarly, males have mechanisms to try and prevent a female from remating. Males might transfer a sticky cement-like substance into the female along with sperm to "plug" the female; in more extreme cases, they might even autotomize (detach) their palp after sperm transfer, such that the detached palp acts as a plug, thus increasing his chances of paternity. If a female uses the transferred sperm to fertilize her eggs, she will lay these and spin a protective egg sac around them until they hatch, typically two to four weeks later. Depending on species, mothers exhibit various levels of maternal care with the eggs and the recently hatched spiderlings. At the very least, the mother will almost always care for the eggs, and often this extends for a period of weeks after hatching. During this entire time, the female will

probably not eat, or have very limited opportunities to do so, since she continuously stays with the eggs and young. This very protection is fundamental for spiderling survival, and once those spiderlings become sexually mature, the cycle begins again.

↑   A male jumping spider will approach the female from the front. He gradually walks astride her as she rotates her abdomen, allowing the male to insert his copulatory organs.

←   A mating pair of orb-weavers. The male is inserting his copulatory organ (the inflated palpal bulb is visible) and transferring sperm into the copulatory ducts on the underside of the female.

THE GOOD MOTHER

Spiders exhibit essentially no paternal care, but females do make especially good mothers: Spitting spiders carry their egg sac in their mouths until they hatch, and wolf spiders carry their spiderlings on their back—sometimes for months after hatching. *Stegodyphus* social spiders (see page 160) not only provide young with regurgitated nutrients, but allow themselves to be fed upon by the spiderlings within the colony, causing death—a behavior morbidly but aptly known as suicidal maternal care. In these instances, *Stegodyphus* show cooperative brood care, regurgitating food and donating their bodies for the care of the young of the entire colony. In fact, of all the hypotheses proposed to explain sociality in spiders, the best supported is extended maternal care, which results in juveniles extending the period for which they remain in the natal web (to adulthood in the most social species)

↑ Many wolf spider mothers will carry their young spiderlings on their backs.

← Other spiders will remain by the nest for the first few weeks of their spiderlings' lives.

and facilitating unusual levels of tolerance between individuals—a trait seldom ascribed to spiders more generally. Some species, such as *Anelosimus vierae*, have taken altruistic feeding to the extreme, with instances of well-fed subadult females regurgitating food to undernourished brothers.

SPERM MATTERS

Spider sex is unusual in another way. The site of sperm production is distant from the copulatory organs, the spoon-shaped palps that males use to inseminate females. Palps are modified legs that are easily observable hanging in front of the spider's mouthparts, while the sperm-producing testes lie within the abdomen connected to the exoskeleton via a small tube. When the male is ready to discharge the level of sperm being stored within his palps, he weaves a small silk mat on which he deposits sperm to move it into storage within the palps in an area called the bulbus.

During mating, he inserts one palp at a time into little copulatory holes or ducts within the epigyne, located on the underside of the female's abdomen. He extends the embolus, and transfers the sperm, along with any plugging material he may have at his disposal. The female can store this sperm for months before fertilizing her eggs when the time is right, such as when she detects warmer weather. Because females that have had a mating plug inserted can often have the plug removed by another male, a female that remates can also store the sperm from multiple males. Little is known about how females selectively choose the sperm from a given male when eventually fertilizing their eggs, but this ability is not only possible, it is increasingly supported by recent studies.

→ Jumping spiders mating. The male's cephalothorax—and palps—are under the abdomen of this female as he inseminates her.

SEX AND THE SPIDER

MALE MATING ORGANS

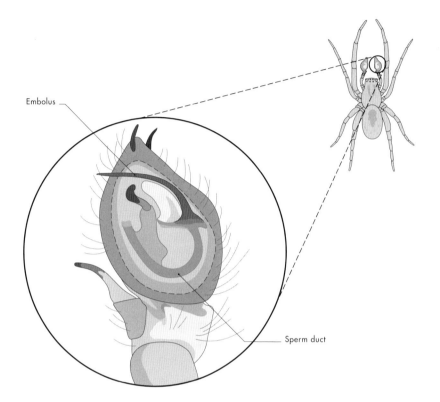

Embolus

Sperm duct

Elaborate palps

Male palp anatomy is complex and highly variable. In fact, it is a key diagnostic feature used for species identification. Sperm stored in the palps is passed down the sperm duct to the embolus, which the male inserts into the female for sperm release.

← This male lynx spider on a leaf is prominently displaying one palp (of two), which is used to transfer sperm to the female copulatory ducts.

→ To maximize the chances of paternity, some male spiders will literally break off the tip of their palp to plug the copulatory duct of the female. This is a scanning electron microscope image showing such an "emasculated" male with both palps broken, leaving him with little stumps.

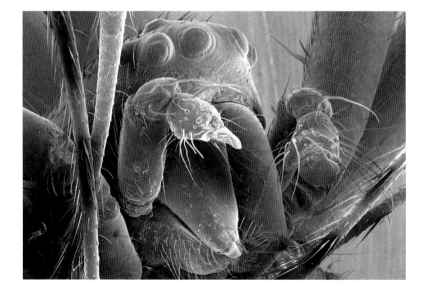

FEMALE MATING ORGANS

Two systems

Spider female genitalia broadly come in two forms. The haplogyne form (left) has a single opening for copulation and egg laying, with sperm, stored in the spermatheca, inseminating the eggs released from the ovaries. The entelegyne form (right) has two copulatory openings, each leading to a spermatheca. Each spermatheca has a fertilization duct leading to the oviduct, which is where eggs are fertilized. Fertilized eggs are then laid through a distinct oviposition opening.

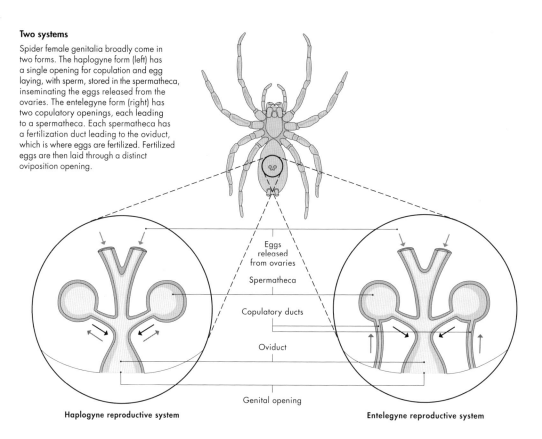

Eggs released from ovaries

Spermatheca

Copulatory ducts

Oviduct

Genital opening

Haplogyne reproductive system

Entelegyne reproductive system

← The copulatory opening of a female *Nephila constricta* with bits of the tips of the palps of three males protruding from a single copulatory duct. Even male emasculation does not guarantee unique paternity!

→ Mating four-spot orb-weaver spiders, *Araneus quadratus*, with the male inserting his palps into the female copulatory ducts to transfer sperm.

Sexual communication

Spider courtship communication ranges from being relatively simple to extraordinarily complex, but it usually involves at least two sensory modalities, often performed in sequence. Many spiders rely primarily on pheromones, which are commonly produced by females and serve as long-distance signals to attract males, and vibratory (seismic) signals, typically produced by males at closer quarters. Males of more visual species also add elements of visual signals in a complex performance, or display.

Displays may involve "dance" movements (motion-based displays), color, and careful ballet-style postures that are held for extended periods—the spider equivalent, perhaps, of a bodybuilder pose. In the final phase, many spiders also have a tactile component to courtship behavior, often involving a male gently tapping or stroking the female with the front pairs of legs and palps as he prepares to mount the female. There are multiple reasons for these elaborate signals, including female assessment of the male's condition as a suitable potential father for her offspring. But the foremost hypotheses include species identification (it pays to mate with the same species!) and the reduction of the chances of a female mistaking the male as potential prey—or simply rejecting him as a prospective mate and attacking and killing him prior to copulation.

← The male *Pisaura mirabilis* will provide the female with a nuptial gift—either a prey wrapped in silk, if he is honest, or just the silken wrapping, if he is less than honest. This male is carrying a gift to entice a gravid (pregnant) female—his chances are slim.

↗ Mating with a giant requires skill, perseverance, and more than a measure of good luck.

LETHAL FEMALES

As voracious predators and typically larger than the males, females pose a significant risk to males, even if they are the same species and even if they are receptive to mating. This is particularly compounded in species in which the male must enter the web—the hunting trap—of the female to court her. While sexual cannibalism (the consumption of potential or actual mating partners) may be especially prevalent in a minority of spider groups, it remains a possibility for all spiders, and it typically involves an attack by the larger spider on the smaller one. Since the male is typically the smaller spider, it is imperative that he assess the willingness of the female to mate before approaching

lethally close to her. In most spiders, sex pheromones present on silk, or sometimes airborne chemicals, allow males to make an initial assessment of female receptivity from a safe distance, and these pheromones can inform the male of the female's size, virgin/mated status, and other handy traits that may promote his survival, such as how well-fed she is.

Having determined that he is in with a chance, a male web-building spider may enter the web and proceed to use vibratory signals on the silk to court the resident female. These closer-range signals may primarily function to reduce the female's urge to attack, as well as to clearly ascertain that both male and female are of the same species. These signals, which can

→ Male jumping spiders (right) typyically perform elaborate courtship dances to woo females.

SPIDER COURTSHIP

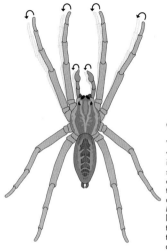

Courtship

Whether on or off the web, spider courtship involves multiple sensory modalties, in- cluding smell or taste, touch and vibrations, and often vision, too. Male spiders use their legs and palps to drum the ground or pluck the web, like a many- armed drummer.

Web-based courtship

On webs, males not only strum the silk with their legs and palps, but may perform whole-body shakes. The precise details of the information that is conveyed during these interactions remains something of a mystery.

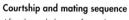

Pre-mating

When a male is near a receptive female, he often carefully taps the female gently before climbing over her for mating.

Mating

Actual mating takes places when the male inserts the embolus from his palp into the female copulatory opening, releasing sperm.

Courtship and mating sequence

After the male has performed an extended sequence of web plucking, drumming, or dancing in front of a female, a willing female will allow the male to walk over her and inseminate her. This does not mean that the male (or, in some cases, even the female) will survive the interaction, though!

be quite varied, may also allow the female to carefully assess the male, which can lead to her allowing the male to transfer more or less sperm during copulation. Male web-based vibratory courtship signals are very different from the vibrations made by struggling prey, in that they are highly repetitive and low amplitude. This may provide the female with a lullaby to her instinct to hunt what is on her web. Signals involving touch, at which point the risk of attack is at its highest, most likely serve a similar appeasing function.

ENTERING THE LAIR

Wandering spiders may not have to enter a female's lair to proceed with courtship, but they, too, must use their signals judiciously for the best possible outcome. In addition to long-range detection via pheromones—often deposited on the silk of draglines deposited by females as they walk through the environment—these spiders typically use either visual or a combination of visual and seismic signals as substrate-borne percussion during courtship.

These visual and seismic elements can be extremely complex, especially among jumping spiders where

seismic and visual signals are often produced simultaneously and synchronously. In these spiders, there may be 20 or more elements within each modality, and these elements are organized into distinct patterns that are produced in sequence, but with variation in the number of repetitions produced for each pattern of elements. In both wolf spiders and jumping spiders, signal complexity seems to be favored because females are more likely to mate with males with complex displays. Since an unreceptive female in these cases can either attack the male or run away, this raises the possibility that showering the female with sensory information may act to overwhelm her into not responding aggressively, or even not at all. However, in many cases, signal complexity is associated with male condition, such that well-fed males are better able to produce long-lasting and complex displays, so there are likely many selective factors favoring complexity among the more visual wandering spiders. Not to be outdone, both wolf spiders and jumping spiders also use tactile signals during the final phase of courtship, which may also reduce female aggression.

SIRING DECISIONS

Because the female is the sex that cares for the young, it is often assumed that female choice drives eventual mating decisions. She invests considerable effort into her young, and so may be rather picky about the males with whom she mates, since the best males might provide good genes for her offspring. Nevertheless, males may not survive their first mating attempt (or maybe not even get that far), so it is not surprising that they, too, might be somewhat choosy about their prospective mates. Males vary the intensity with which they court females, based not only on their own body condition or size, but also on their assessment of the likely receptivity of the female, the risk posed by the female, and the risk of another male mate shortly afterward potentially siring most of the offspring.

← A male wolf spider extending his palp down under the female to mate.

Sexual size dimorphism

Many terrestrial animals, including humans, display differences in size between the sexes, but nowhere is this more prevalent or extreme than in spiders, where cases of sexual size dimorphism (male and female morphology) typically involve females as the larger sex. While females can often be three to more than ten times larger than the males, in cases where males are larger, this is only by a small fraction.

Spiders grow by molting, and th᾿ size change before and after the molt depends on genetics, how well fed the spider is, and other developmental factors that demonstrate considerable plasticity in the dynamics of spider growth. These effects result not just in larger females, but also in considerable variation in male body size within a given species. Experiments on *Nephila cruentata* suggest that adult male size is in part driven by perceived future male–male competition during their subadult period. Subadult males exposed to silk-based cues from adult male *N. cruentata* molt to become larger than those exposed to cues from adult females. A further contributor to final size is the number of molts. In species in which females are very large, females have a shorter period between molts compared with males, and thus molt more frequently before reaching their final molt at sexual maturity (although in rare cases, molting continues in sexually mature spiders, such as in females of the golden orb-weaver *Nephila pilipes*).

→   The size difference between the male *Argiope aurantia* and the female highlights the magnitude of the problem the male faces to mate with an often voracious—and always predatory—female.

↓   The best way for a male to guarantee the virginity of a female is to wait for a subadult female to molt into her last, sexually mature, developmental stage (known as instar). This is the shed exoskeleton of an orb-weaver.

GREATER FECUNDITY

Larger female spiders are more fecund: They can lay more and larger eggs that are more likely to hatch, so selection for overly large females might be due to males choosing them as mates. However, this does not fully explain the extreme sexual size differences prevalent in many spiders. For example, extreme sexual differences in size are especially common among web spiders, where females are largely sedentary. In some of these cases, size dimorphism is likely a consequence of female gigantism—in which females molt more often and become larger—rather than due to male dwarfism, whereby selection acts on males to remain small. Selection for small males could occur, for example, because they may sneak copulations with inattentive females (possibly guarded by larger inattentive males), or they may be deemed too small to cannibalize.

However, the drivers of extreme sexual size dimorphism do not fit one pattern: In some instances, although males may benefit by being as small as possible, larger males may be better placed to guard females, which may lead to less extreme dimorphism between the sexes. Then again, as males actively search for the sedentary females, they are exposed to high predation risk, which may be more pronounced among larger males, possibly selecting for smaller males. Numerous other suggestions have been proposed to explain extreme sex size dimorphism in spiders, but one size does not fit all (pun intended), and the evolution of sexual dimorphism—particularly in its extreme form—is still poorly understood. What is probable is that different groups of spiders will face different selective pressures, and so a single unifying explanation is unlikely to occur.

A QUESTION OF SIZE

The extreme female-biased sexual size dimorphism in spiders is related to numerous intriguing aspects of spider sexual behavior, including a male-biased sex ratio (more males than available females in a population at any given time), leading to high levels of competition between males, sexual cannibalism, and male emasculation, whereby the male loses his copulatory organ during sex. A simple thought experiment suffices for considering the risk for a male seeking to mate with a voraciously predatory female several times his size: If an average c. 6 foot (182 cm) human male weighing 176 lb (80 kg) were in the position of many nephilid spiders, he'd *very carefully* consider how to approach the aggressive 33-foot- (10-meter-) tall female weighing around a ton!

← A spider sheds its "skin." The only way for spiders with an exoskeleton to grow is to molt. This leaves them vulnerable while they harden into their new cuticle and also means that they are unable to attack any males that may be looking to mate.

Sex is risky

Polyandry, or females mating multiple times with different males, is common in spiders. Paternity patterns are variable, with first male sperm precedence for egg fertilization in some species, last male sperm precedence in others, and in yet other species, females have strategies that enable them to pick sperm selectively from a given male from within the reproductive tract and use this sperm for egg fertilization. The latter is known as cryptic female choice.

It is likely that females of several spiders that mate with multiple males can bias the paternity of their offspring through cryptic female choice, either by selectively dumping sperm from unwanted males, as found in the daddy longlegs spiders (Pholcididae) *Pholcus phalangioides* (see page 190) and *Physocyclus globosus;* the long-jawed orb weaver (Tetragnathidae) *Pachygnatha clercki;* and the goblin spider (Oonopidae) *Silhouettella loricatula*. Alternatively, this can be accomplished by selectively storing sperm from preferred males, as found in the nursery web spider *Pisaura mirabilis* and the orb weaver *Argiope lobata*.

PATERNITY STRATEGIES

Consequently, males have evolved many strategies to increase their chances of paternity. Mate guarding is especially common in web spiders, where several males may loiter at the extremities of a female's web, waiting for an opportune moment to approach. If an impending sexually mature female is resident (which is detected by pheromones), a male may guard the female and aggressively spurn the approach of other suitors until he has mated with the female when she reaches sexual maturity and even for some time afterward, increasing the chances that any laid eggs will be his. Other strategies used include providing "nuptial gifts," plugging female genitalia, and self-sacrifice.

The provision of nuptial gifts, where males offer some form of nutrients to females before, during, or directly after copulation, is rare in spiders—it occurs only in some species of pisaurid nursery web spiders and trechaleid fishing spiders. This has been most studied in *P. mirabilis*, where gift giving dramatically increases the male's chance of mating. Not only that,

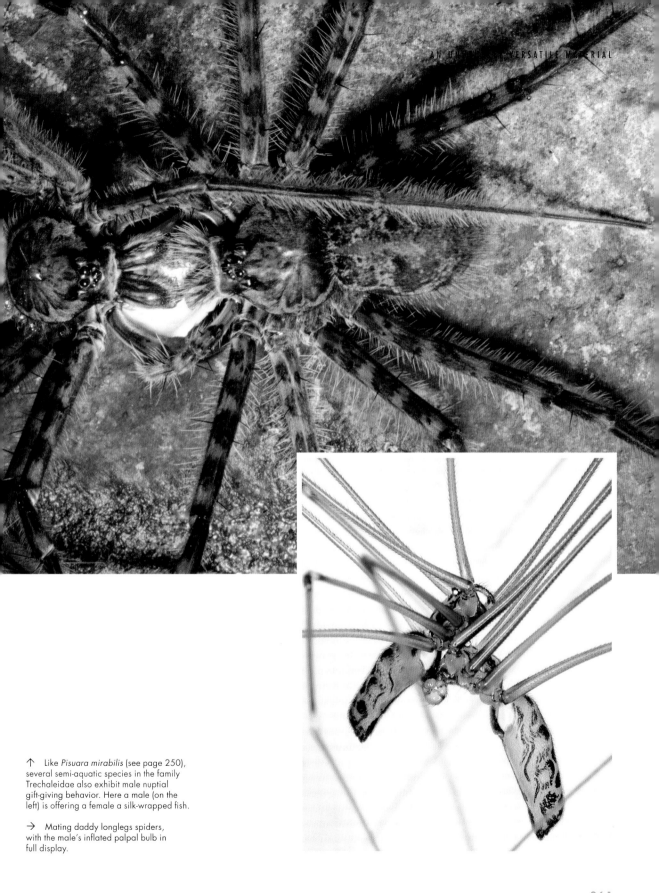

↑   Like *Pisuara mirabilis* (see page 250), several semi-aquatic species in the family Trechaleidae also exhibit male nuptial gift-giving behavior. Here a male (on the left) is offering a female a silk-wrapped fish.

→   Mating daddy longlegs spiders, with the male's inflated palpal bulb in full display.

but when copulation duration is experimentally matched between males offering no gifts and gift-giving males, more of the sperm of generous males is stored by the female through cryptic female choice. This suggests that females assess the males with which they mate, and depending on that assessment, they store more sperm from males deemed worthy, as seen in the orb-weaver *Argiope bruennichi*, where males that mate with females without bothering to court the female beforehand have a reduced paternity share of the fertilized eggs than males that adopted courtship behaviors prior to mating.

TOP-NOTCH GIFTS

Gift-giving males of *P. mirabilis* offer females neatly silk-wrapped prey items during courtship or mating. The female proceeds to eat this gift, and the male is able to mate with her. Yet, as in any system, a male may use deception, and the gift may simply be silk without prey inside or have inedible items like seeds hiding beneath the wrapping. However, females cotton on to the underwhelming gift quickly, and copulation duration is shorter when females are not provided with nutritious food. This may lead to reduced sperm transfer during copulation, and consequently it is likely they will get smaller chances of paternity, so cheating males don't get all the goods. In turn, this suggests that honest males providing females with substantial prey items (those that will keep her busy for some time)

will be able to transfer large amounts of sperm and increase their chances of paternity.

A more common strategy used by males is the use of mating plugs, whereby males try to secure paternity by plugging female genitalia to reduce another male's chances of transferring sperm. In some instances, female secretions are combined with male secretions to produce a mating plug. While producing secretions to plug female genitalia may not prevent female remating—in many cases subsequent males can remove the plug—it certainly does not prevent the male remating. In stark contrast, plugs in which males break off part of their palp to leave within the female certainly do influence their future reproductive success. In the best-case scenario, the emasculated male would have two mating opportunities under this strategy, and male emasculation is especially common in species with extreme sexual dimorphism and limited chances for a male to secure a future mating. The golden orb-weaver *Nephila fenestrata* exemplifies the issue. Breaking off a section of their palp to provide a plug during mating, males are rendered functionally sterile. However, subsequent males can remove the plug, so it pays for the male to guard the female from other suiters, even when there is considerable risk of further injury by doing so. He has nothing more to gain in terms of future paternity, so he will fight for the paternity he has obtained.

← Males will often take up residence in a female's web, guarding her until she molts into sexual maturity or, having mated, fending off other suiters to maximize his chances at paternity. Note also the zig-zag stabilimentum on this web.

CANNIBALISM

Even the cost of losing the copulatory organ by a male spider
or injury pales in comparison to losing his life and becoming his
mate's next meal. Sexual cannibalism is particularly common
among spiders with extreme differences in size between the larger
female and the smaller male, as found among many orb-weavers.
For example, up to 80 percent of male *Argiope bruennichi* will be
cannibalized during their first mating attempt, and the chances
of being killed by the female increase to 100 percent with longer
mating durations. Since longer mating also increases the male's
paternity, the investment of his life is worthwhile—at least in
many instances.

Males are discriminatory in whether they will volunteer their
life, for example, depending on their compatibility with the female.

↑   Mating nursery web spiders,
Pisaura mirabilis. The inflated palpal
bulb of the male, containing the sperm
to transfer to the female copulatory
ducts, is visible.

↗   Here, the female jumping spider
(below the male) has just molted into
sexual maturity. This presents a perfect
chance for a male to mate and secure
his chances at paternity.

Males that mate with siblings copulate for very short periods, increasing their chances of survival compared with males that mate with unrelated females that are more compatible. Other mechanisms used by this species—and many others—to increase survival include mating with molting females they have been guarding in anticipation of their sexual maturity. Having a soft exoskeleton, a molting female is unable to attack the male. Male *A. bruennichi* that mate with molting females reduce their mortality to near zero—a very effective strategy—and since the female has only just sexually matured, she is also a virgin. Should these males successfully insert a mating plug, they have a reasonable chance of paternity and even live to try again with another female. It may seem that this strategy is so effective that there should be no other, but females in a given population typically reach sexual maturity at about the same time, so a male guarding a subadult female for several days may lose opportunities to mate with early maturing females—and having mated with a molting female, his chances of finding another subadult to guard become vanishingly small. The spider sex tug-of-war is fraught with risk, and the vested interests of females and males sometimes work antagonistically, resulting in this kaleidoscope of sexual strategies.

Golden silk orb-weaver

Male eater

SCIENTIFIC NAME	*Trichonephila clavipes*
FAMILY	Araneidae
BODY LENGTH	Females 1–1⅗ in (25–40 mm), males ¼ in (6 mm)
NOTABLE ANATOMY	Females are large orange and brown spiders with feathery tufts on the legs
MEMORABLE FEATURE	May store prey in larders on the web for future consumption

There is a widespread idea that for many animals, males should mate with as many females as possible, while females should be choosy because of the costs they incur by mating. This oversimplification assumes there are no costs to males, but there are. Males need to search, possibly fight for, may be killed by female mates, and they may have limited sperm supplies to name but a few. These—often extreme—costs are incurred by males in the orb-weaver subfamily Nephilinae, famous because they exhibit the most extreme sexual size dimorphism of any terrestrial animal.

GIANT FEMALES

Nephiline females can be six times larger and can weigh 20 times more than the males. This is not male dwarfism (a reduction in the size of males through evolutionary time), but is instead female gigantism. To achieve their great size, females have a longer developmental period with more instars, requiring a more generous diet than males to reach maturity. These discrepancies make this an excellent group for studying differential aspects of selection on males and females. Strong selection has led to female gigantism, possibly due to reduced predation and increased ability to produce more offspring. However, while there is considerable variation in male size, males, too, tend to be large—they just appear minute by comparison with the females.

MALE COSTS

On reaching sexual maturity, the male leaves his natal web and searches for a female. These searches are dangerous: A male *Trichonephila* may have only 25 percent chance of surviving just this phase of his reproductive effort. Having made it to a female's web, the nephiline male will establish himself, often with many other males residing on the web, and fight for access to the female. Here, larger males have an advantage, since they can outcompete other males in contests for the female, but in sexually cannibalistic species such as *T. clavipes*, this is a double-edged sword, as larger males are more likely to be cannibalized, at least by virgin females.

Female *T. clavipes* can mate multiple times; so to increase chances of paternity, a male will often guard the mated female—but his motivation to do so is strongly affected by how many competitors he has on the web. In other words, a male might trade the opportunity to search for, and mate with, another female in return for maximizing the returns on the current mating investment. However, with about a 50 percent chance of being cannibalized by the female, postcopulatory mate guarding is often a moot issue.

→ When cutting their silk lines, *Trichonephila clavipes* bring the line to their mouthparts but do not mechanically cut the silk. Instead, it seems they use an enzyme that cuts silk on contact.

Dark fishing spider

Martyr males

SCIENTIFIC NAME	*Dolomedes tenebrosus*
FAMILY	Pisauridae
BODY LENGTH	Females ⅗–1 in (15–25 mm), males ⅓–c. ½ in (7–13 mm)
NOTABLE ANATOMY	Brownish-gray flecked with black and lighter brown markings; banded legs
MEMORABLE FEATURE	Mating always results in male death immediately upon insemation of female

Several theories have been proposed to explain sexual cannibalism of males by females during or shortly after copulation. For example, by sacrificing themselves, a male may decrease the chance of the female remating, or he may provide her with nutrients that will be favorable for her laying eggs and promoting the survival of young. *Dolomedes* fishing spiders (Pisauridae) often engage in sexual cannibalism; in some cases, all males sacrifice themselves, meaning that they get one shot at paternity. More problematic is precopulatory sexual cannibalism, as sometimes seen in *D. tenebrosus*, where the male is consumed *before* insemination.

TO EAT OR NOT TO EAT

One intriguing hypothesis that seeks to explain precopulatory sexual cannibalism is the aggressive spillover hypothesis, which posits that females that are extra-gluttonous as juveniles gain survival advantages. This aggressive personality then nonadaptively spills over into the sexual domain as an adult. This also entails costs to the females. After all, a female's mating success is considerably impaired if she eats all males that attempt to mate with her. There is some evidence to support the spillover hypothesis, but precopulatory sexual cannibalism is also related to either the number of previous mates a female has had or the size discrepancy between males and females, with larger males tending to evade a lethal outcome. As yet, this is an unresolved enigma.

MONOGAMOUS MALES

D. tenebrosus exhibits male monogamy, where males only mate once. In fact, *D. tenebrosus* self-sacrifice—they spontaneously die and are consumed by the female directly after mating. Many factors may favor male monogamy: Males mature early to increase their chances of finding a female, there is a male-biased sex ratio (and thus competition between males for access to females), and there is sexual size dimorphism. Nevertheless, since the much larger females readily remate, male paternity is far from guaranteed, leading to questions regarding the adaptive advantage of self-sacrifice during the mating sequence. However, there are advantages: Being eaten by the female substantially increases the number, size, and survivorship of the ensuing offspring. This is independent of the female simply obtaining food. When offered a cricket as an alternative food source, the number and mass of offspring, as well as their survival, is as poor as if the female consumed nothing after copulation. In contrast, when she eats the male, spiderling mass, number, and survival almost double. Possibly, the male provides the female with key nutrients favorable to offspring survival.

→ Among the species least affiliated with water among the *Dolomedes* genus, *D. tenebrosus* can be found in swamps and ponds but also in terrestrial habitats.

Running crab spider

Promiscuous partner

SCIENTIFIC NAME	*Philodromus cespitum*
FAMILY	Philodromidae
BODY LENGTH	Females ⅕ in (5 mm), males ⅛–⅕ in (3.5–5 mm)
NOTABLE ANATOMY	Variable brown or yellowish-brown body color, mostly spotted
MEMORABLE FEATURE	Ambush predators that often rest with legs splayed to the side

In spiders, less draconian measures than suicide-by-sex exist to increase paternity. One is the frequent use of mating plugs, or the application of mechanical barriers to the female copulatory openings (which differ from the oviposition openings in most species) to prevent future insemination by other males.

Plugs can comprise broken male genitalia left inside the female, which come at a cost to the male's future mating success. Alternatively, they can be in the form of quickly hardening glandular secretions produced either by the male or, in some cases, by both sexes. A female does have agency in this process: She can exert choice about which males she mates with and the effectiveness of a given plug insertion, and she can also store sperm from previous copulations to fertilize eggs. In turn, males can often remove the plugs from previous suitors.

WHO IS CHOOSY?

Commonly found on the leaves of fruit trees in European orchards, where it acts as a natural enemy against pest species, *Philodromus cespitum* is a promiscuous, secretion-plug-producing spider. Males always insert a plug when they transfer sperm, although this does not prevent females from remating. Possibly because the plugs of *P. cespitum* are not exceptionally successful at protecting paternity, females are not choosy about their mating partners. Additionally, females terminate copulation, and this likely influences the amount of plug material deposited by the male. Nonetheless, males are attracted to virgin females, suggesting that the plug provides some success in siring offspring. Males can discriminate between juveniles and adult females, and between virgin and mated females. Among mated females, they can also discriminate those with small plugs compared to large plugs based solely on chemotactile cues on their silk draglines.

MALE APPENDAGES

The copulatory organs of male spiders are modified legs that hang in front of the mouthparts (also present without a reproductive function in females). Because the palps are physically separated from the site of sperm production, sperm is taken into and stored within the palpal bulbi for eventual transfer to the female. Until recently, it was thought that spider male genitalia were not innervated and thus lacked any sensory input. However, in 2017, an examination of *P. cespitum* revealed many sensory neurons within the palps and associated with the palpal bulbi. These have now been found in all other spider taxa investigated, including mygalomorphs, suggesting that nervous tissue is likely present in all species, which until now has been simply unobserved.

→   The use of insecticides for orchard pest control has negative sub-lethal effects on the predatory behavior of running crab spiders.

Paradise spider

Sexy drummer

SCIENTIFIC NAME	*Habronattus pugillis*
FAMILY	Salticidae
BODY LENGTH	⅕–⅓ in (5–7 mm)
NOTABLE ANATOMY	Highly variable coloration and markings based on geographical population
MEMORABLE FEATURE	Live at high elevations; males produce synchronized visual and seismic courtship displays

The courtship displays of *Habronattus* jumping spider species are among the most complex seen in arthropods, and variation in displays within a species can be thought of as dialects. In *Habronattus pugillis*, differences in courtship dialect may be leading to the formation of different species. The independent mountain ranges in Arizona, USA, and Northern Mexico are home to unique populations of *H. pugillis*.

These populations, even when only geographically separated by a few miles, exhibit diversity in male appearance and courtship behavior. While at least some populations are capable of interbreeding, the populations remain distinct because each makes its home on a separate "sky island" mountain range, and there is strong local female selection for distinct male traits.

SPECIATION

Females of *H. pugillis* exert strong selection on males, where variation is especially apparent. While there are common elements of male courtship behavior across populations, each population has distinct, unique visual and vibrational courtship signals. This signaling variation matches variation in male coloration. Males are brown, silver-gray, and black, with elegant splashes of white, but the markings differ between populations. Specifically, it is the markings on the face, first pair of legs, and palps—elements used in courtship displays—that show the most interpopulation variation. Crucially, because divergence between populations is

happening so rapidly, the formation of separate species— speciation—can be documented as it is happening, as some between-range crosses already lead to unusually low numbers of offspring.

IN-GROUPS, OUT-GROUPS

Females from some populations are wary of outsiders and slow to accept males from other populations as mates. When they do mate with these foreign males, the females produce few eggs. However, other cross-population pairings remain viable. In some cases, females seem to be more attracted to males from other populations than their own local males. Females from the Santa Rita mountains, for example, are much more likely to mate, and mate faster, with males from the Atascosa Mountains than their own Santa Rita males.

However, the reverse is not true: Atascosa females behave similarly toward Atascosa and Santa Rita males; they don't prefer the foreign males, but they don't avoid them either. Santa Rita female preference for Atascosa males, however, is abolished if the vibratory display—more complex in the Atascosa population—is experimentally muted, showing that it is the vibratory courtship component that drives this Santa Rita female preference for the out-group males.

→   These oak forest montane spiders have a very broad tolerance for a wide range of temperatures.

Brush-legged wolf spider

Cunning rival

SCIENTIFIC NAME	*Schizocosa ocreata*
FAMILY	Lycosidae
BODY LENGTH	¼–⅖ in (6–10 mm)
NOTABLE ANATOMY	Distinct pale longitudinal band down entire body
MEMORABLE FEATURE	Ground-dwelling spider that varies its foraging patch residence time

Female *Schizocosa ocreata* typically mate once only, while males mate multiple times, creating competition for females. This is reflected in early male maturation, in which males are sexually mature and ready when females begin to mature in the spring. Roving around the leaf litter looking for females, males respond to pheromones on silk to detect the presence of a suitable unmated adult female, preferring these to the pheromones from mated females, which are more likely to cannibalize males.

HIDDEN MEANING

On locating a female, males begin a courtship display in which they use a combination of seismic signals and visual displays showing off the brushes on their first pair of legs to persuade the female to mate. Both visual and seismic signals provide information about the males' condition and quality, such as his size, feeding history, and aerobic fitness. However, the use of multimodal signals enhances detection and elicits a quicker and more positive female response compared with signaling in either one of these modalities. There are other advantages to having a backup communication channel. *S. ocreata* live on ground covered by leaf litter, rocks, and bark; some of these, such as leaf litter, are better than others, such as rocks or soil, at transmitting vibrational information.

When males are on soil or rocks, they cleverly use more visual signals then seismic signals. Visual signals are detectable from twice the distance of seismic signals, but while both males and females prefer leaf litter, leaves may also obstruct the view of the waving legs of a ground-dwelling spider, and the signal may fail to reach the intended female. In the dark, seismic communication can also convey information where visual signaling is ineffective. So, while each modality may convey the same information, the spiders use each adaptively to ensure maximal transmissibility in each environment.

SUBTLE ART OF EAVESDROPPING

Although multimodal signaling enhances detection of the males by the female, it also does so among other males, which may be alerted to the presence of a nearby female by the displays of another male. What do they do when they detect a courting male? Begin courting … after all, male mating success is determined by the number of females he encounters and can mate with, and any trick in the book goes in finding a female. However, unlike females that respond equally to each sensory modality, males appear to weigh seismic information more heavily than visual information, perhaps because, as a short-range signal modality, this likely indicates that the unseen female is very close.

→   The loss of a leg is common among brush-legged wolf spiders. As male displays rely on their leg brushes, this will mean reduced courtship success, even if limbs are regenerating—females like symmetry.

MARATUS VOLANS

Peacock spider

Magic dancer

SCIENTIFIC NAME	*Maratus volans*
FAMILY	Salticidae
BODY LENGTH	Males ⅛–⅕ in (4–5 mm)
NOTABLE ANATOMY	Males display iridescent yellow, green, orange, and opalescent blue "flaps," which are extended during courtship behavior
MEMORABLE FEATURE	Complex multimodal courtship display behavior by males

Writing in the *Journal of Performance Art* about the seismic courtship signals of *Maratus volans* jumping spiders, the composer and professor David Rotherberg states that: "The even spacing of the beats shows … rhythm and … each sound has a unique sense of tone, not a simple click or hit. The presence of patterns in the midst of clouds of noise shows that there is a glimmer of music inside the spider's careful motions." True enough: If any spider were to be lauded as a performance artist, it would have to be this one.

ARTFUL COMMUNICATION

Male *M. volans*, like other *Maratus* species, are justly renowned for having among the most varied and complex courtship displays known of any animal. Elaborate motion-based visual displays in which the legs are waved in time with the artful presentation of a slew of red, blue, green, and orange colors on swivelling folds that unfurl like fans from their abdomen form the visual component of the display. The "music" of the dance, produced by abdominal fluttering, is added to this.

EYE OF THE BEHOLDER

How does an observing female react to this performance? Well, if she is not impressed, she lifts and wiggles her abdomen, signaling her disinterest. Until that happens, the male's aim is to get the female's rapt attention; if she turns and looks away, he changes rapidly from the visual dance, to "singing" through seismic signals. Once her attention is drawn back to the male, he once again pivots to the visual elements of his courtship, since it is these—and the vigor with which he shakes his colorful peacock-like fans—that seem key to persuading the female to mate. The signaling modalities appear to play complementary roles: Visual signals keep the female's eyes locked on the male, and seismic signals get her attention if she is not looking at him.

However, despite the beauty of the colors on display, it seems likely that color plays a less important role in these displays than the contrast between colors or the pattern of brighter and darker shapes. That is not to say that color *per se* plays no role in her decision but it seems the male's colors do not have the perceptual saliency they do in our visual system.

Mating dance

During the extravagant visual courtship dance of *Maratus volans* the male will raise his third pair of legs vertically and lower them to the side, while extending flaps on his near vertically raised abdomen. He then rapidly bobs this side to side, while continuing the leg movements and approaching the female.

→ If any spider can be said to have gone "viral" on the Internet, the peacock spider is the one!

GLOSSARY

↑ The sensory organs of the tiger bromeliad spider, *Cupiennius salei* (see page 56), on full display. Note the characteristic positioning of the eyes, the dense covering of short tactile hairs (like fur), and, amid the dark spines on its legs, the longer thicker trichobothria hairs that deflect with the tiniest breath of wind.

aciniform silk Silk produced by the aciniform gland that is typically used to wrap prey or line eggsacs.

ampullate glands The glands (major and minor) within the spider that produce ampullate silk, which is used to create the frame of a web (major ampullate silk), and for the radial threads of a web (minor ampullate silk).

aposematic Defensive bright coloration, signaling distastefulness.

arachnologist Scientist that studies spiders, ticks, scorpions, or mites (arachnids).

Araneomorph Spider infraorder comprising over 90 percent of species, characterized by the more "modern" placement of the chelicerae pointing diagonally forward.

arcuate body A region of the spider's brain involved in higher-order processing of information, likely integrating information received through multiple sensory inputs.

Batesian mimicry When a harmless, edible species resembles a distasteful or harmful, often more common, species (such as an ant), thus gaining some protection from predators.

Beltian body A detachable structure, rich in lipids and proteins, that is found at the tips of the leaflets of some plants, such as some acacia.

bioinsecticide A pesticide that works against one or more insects and which is not synthetically but naturally derived.

book lung Respiratory organ found in some spiders and in scorpions. It consists of a series of parallel tissues (lamellae) that are well supplied with blood (hemolymph) to provide gas exchange.

cephalothorax The fused head and thorax of chelicerate arthropods, such as spiders.

chelicera The mouthparts of spiders and other arachnids, horseshoe crabs, and sea spiders (subphylum Chelicerata).

chemoreception The ability to detect chemical stimuli. Can include the senses of taste and smell.

chemotactile The ability to detect chemical stimuli through direct contact (touch), as in the sense of taste.

conspecific Members of the same species.

crypsis Colored to blend in with the substrate, as opposed to aposematic.

cuticle Epidermis or skin of a spider or insect.

cylindriform silk Tough silk produced by the cylindriform (also known as tubiliform) gland that is used to coat eggsacs for protection.

exoskeleton The cuticle of a spider that protects and supports its body, and to which its internal muscles attach.

extensor muscle Muscle that, when contracted, straightens (extends) the limb or body part.

extraoral digestion The secretion of enzyme-rich fluids to dissolve prey tissue before digestive processing in the gut.

flagelliform gland The gland within the spider that produces flagelliform silk, which is used to create the spirals within a web.

flexion Bending (e.g., of a limb).

gigabase (Gb) Measurement unit that designates the length of DNA molecules. One Gb is one billion bases, or nucleotides (the basic structural unit of DNA).

hymenopteran An insect belonging to the order Hymenoptera, which includes ants, bees, and wasps, among others.

liphistiid An ancient family of spiders consisting of about 140 species.

longitudinal waves Waves in which the medium (e.g., silk) vibrates in the direction of their propagation.

loxocelism A condition resulting from the bite of spiders in the genus *Loxosceles*.

lyriform organs A term used to denote a cluster of slit sensilla.

mechanoreception The sensory ability pertaining to perception of mechanical stimuli, such as pressure, deflection, or distortion

metabolic rate The energy expended by an animal in a given period of time.

nectarivory A type of foraging in an animal that derives its energy and nutrient requirements from a diet consisting primarily of plant nectar.

New World North, Central, and South America.

Old World Africa, Europe, and Asia, or Afro-Eurasia.

orb-weaver Common name for spiders in the family Araneidae.

orb web Circular web shape often constructed by orb-weaving spiders.

oviposition Laying of eggs.

palps An abbreviated form of the word "pedipalps" (see below).

palpal bulbi Plural of palpal bulb, the copulatory organ of male spiders.

pedipalps In chelicerates, the pedipalps (or palps) are the second pair of appendages. They lie alongside the chelicerae, or mouthparts, and in front of the legs.

pheromone Chemical substance that is secreted by an animal into the environment and which can be used for communication.

photoreceptors Specialized cells of the nervous system that convert light (captured photons) into signals that can be used to guide behavior.

riparian Relating to areas near a natural watercourse (e.g., rivers and streams). A riparian zone is at the interface between the water source and land.

theridiids Family of spiders known as tangle-web spiders, cobweb spiders, gum-footed spiders, and comb-footed spiders.

thermoregulation Mechanism used by animals to maintain a relatively steady body temperature independent of the external environmental temperature.

parasitoid A parasite (e.g., a wasp or fly) whose larvae feed on a host (such as a spider), developing inside (sometimes outside) the host and eventually killing it.

phylogenomics The study of evolutionary history that uses genomic techniques to determine the relationships between different groups, or taxa, of organisms.

phylogenetics The study of phylogeny.

phylogeny The representation of the evolutionary history of a group of organisms depicting relationships between different groups, or taxa.

planar web A two-dimensional web that lies on a plane (e.g., vertical, horizontal, or angled).

predation When an animal actively expends energy to obtain another organism (which puts up some resistance to being predated on) for nutritional gain.

proprioception The sense that enables an animal to perceive the body's location, movement, and limb or joint positioning within space.

pyriform silk Silk produced by the pyriform gland that is used to attach webs to the substrate.

rhabdomeres parts of a rhabdom, which are the light-sensitive rod-like structures within arthropod photoreceptors.

sensillim, sensilla (pl) Microscopic organs of arthropods, used to sense the world (e.g., taste, smell).

slit sensilla Also known as the slit sense organ. Comprised of one or more minute "slits" in the spider's cuticle, which detect physical deformation, or strain.

sociality The propensity of individuals within a species to associate with each other or live in groups.

spinneret In spiders and other insects, an organ for spinning silk that is secreted as liquid by silk glands and usually solidifies to become a silk thread.

subesophageal ganglion Region of the spider's brain that is located below the esophagus.

subphylum A biological taxonomic classification that is below a phylum.

substrate An underlying layer; a surface on which an animal lives.

supraesophageal ganglion Region of the spider's brain that is located above the esophagus.

tapetum (plural tapeta) A layer of reflecting cells found in the eyes of many (particularly nocturnal) animals. By reflecting light back through the retina, tapeta increase the light available to photoreceptors.

transverse waves Waves in which the medium (e.g., silk) vibrates at right angles to the direction of their propagation.

trichobothria Stiff articulated hairs present on a spider's or insect's body, which are used to detect air currents and electrical charge in the environment.

troglobiont An animal that lives in caves.

tubiliform silk Spider silk produced by tubiliform, or cylindriform, glands.

vibrometry Pertaining to the use of a laser vibrometer to measure tiny oscillations in a medium.

zodariid Family of spiders known both for hunting ants and for having a slight resemblance to ants.

RESOURCES

BOOKS

Barth, F.G. (Ed.). *Neurobiology of Arachnids*
(Berlin; New York: Springer-Verlag, 1985)

Foelix, R.F. *Biology of Spiders*. 3rd edition
(New York: Oxford University Press, 2011)

Herberstein, M.E. (Ed.).
Spider Behaviour: Flexibility and Versatility
(Cambridge: Cambridge University Press, 2011)

Nentwig, W., J. Ansorg, A. Bolzern, H. Frick,
A.S. Ganske, A. Hänggi, C. Kropf, and A. Stäubli.
All You Need to Know About Spiders.
(Cham: Springer Nature, 2022)

BOOK CHAPTERS

Harland, D.P., D. Li, and R.R. Jackson. "How jumping
spiders see the world." In O.F. Lazareva, T. Shimizu,
and E.A. Wasserman (Eds.). *How Animals See the
World: Comparative Behavior, Biology, and Evolution
of Vision* (New York: Oxford University Press, 2012,
pp. 133–164)

Herberstein, M.E. (Ed.). "Plasticity, learning and
cognition." In *Spider Behaviour: Flexibility and Versatility*
(Cambridge: Cambridge University Press, 2011)

SCIENTIFIC JOURNAL ARTICLES

Aguilar-Arguello, S., and X.J. Nelson. "Jumping spiders:
An exceptional group for comparative cognition
studies." *Learning and Behavior*, 49 (3): 276–291 (2021).
https://doi.org/10.3758/s13420-020-00445-2

Barth, F.G. "Spider senses: Technical perfection and
biology." *Zoology*, 105 (4): 271–285 (2002). https://doi.
org/10.1078/0944-2006-00082

Gaffin, D.D., and C.M. Curry. "Arachnid navigation:
A review of classic and emerging models." *The Journal
of Arachnology*, 48 (1): 1–25 (2020). https://doi.org/
10.1636/0161-8202-48.1.1

Japyassú, H.F., and K.N. Laland. "Extended spider
cognition." *Animal Cognition*, 20 (3): 375–395 (2017).
https://doi.org/10.1007/s10071-017-1069-7

Liedtke, J., and J.M. Schneider, "Association and reversal
learning abilities in a jumping spider." *Behavioral
Processes*, 103: 192–198 (2014). https://doi.org/
10.1016/j.beproc.2013.12.015

Meehan, C.J., E.J. Olson, M.W. Reudink, T.K. Kyser,
and R.L. Curry. "Herbivory in a spider through
exploitation of an ant–plant mutualism." *Current
Biology*, 19 (19), R892–R893 (2009). https://doi.
org/10.1016/j.cub.2009.08.049

Morehouse, N. "Spider vision." *Current Biology*, 30 (17):
R975–R980 (2020). https://doi.org/10.1016/j.
cub.2020.07.042

Pekár, S., S. Toft, M. Hruskova, and D. Mayntz. "Dietary
and prey-capture adaptations by which *Zodarion
germanicum*, an ant-eating spider (Araneae : Zodariidae),
specialises on the Formicinae." *Naturwissenschaften*,
95 (3): 233–239 (2008). https://doi.org/10.1007/
s00114-007-0322-3

← Near-blind web-building
spiders are nevertheless
awe-inspiring predators.

→ Large, ground-based spiders,
are a marvel to behold as they go
about their lives—which can be
over 30 years.

Rößler, D.C., K. Kim, M. De Agrò, A. Jordan, C.G. Galizia, and P.S. Shamble. "Regularly occurring bouts of retinal movements suggest an REM sleep–like state in jumping spiders." *Proceedings of the National Academy of Sciences*, 119 (33): e2204754119 (2022). https://doi.org/10.1073/pnas.2204754119

Selden, P.A., and D. Penney. "Fossil spiders." *Biological Reviews*, 85 (1): 171–206 (2010). https://doi.org/10.1111/j.1469-185X.2009.00099.x

Stafstrom, J.A., G. Menda, E.I Nitzany, E.A. Hebets, and R.R. Hoy. "Ogre-faced, net-casting spiders use auditory cues to detect airborne prey." *Current Biology*, 30 (24): 5033–5039 (2020). https://doi.org/10.1016/j.cub.2020.09.048

Wise, D.H. "Cannibalism, food limitation, intraspecific competition and the regulation of spider populations." *Annual Review of Entomology*, 51: 441–465 (2006). https://doi.org/10.1146/annurev.ento.51.110104.150947

Zurek, D.B., T.W. Cronin, L.A. Taylor, K. Byrne, M.L.G. Sullivan, and N.I. Morehouse "Spectral filtering enables trichromatic vision in colorful jumping spiders." *Current Biology*, 25 (10): R403–R404 (2015). https://doi.org/10.1016/j.cub.2015.03.033

ARACHNOLOGY SOCIETIES AND USEFUL WEBSITES

American Arachnological Society
www.americanarachnology.org

British Arachnological Society
https://britishspiders.org.uk

European Society of Arachnology
www.european-arachnology.org

International Society of Arachnology
https://arachnology.org

World Spider Catalog, Version 24.5
https://wsc.nmbe.ch rg/wiki/Lepidoptera

INDEX

ACKNOWLEDGMENTS

I would like to thank Tom Broadbent, Kathleen Steeden, Wayne Blades, and Sarah Skeate, as well as all the other helpful people at UniPress. Their work has really helped bring the text to life. I am also grateful to Matjaž Kuntner, for the use of several beautiful SEM pictures, and Jeremy Squire, Lou Staunton, Simon D. Pollard, Marshall Hedin, Jurgen Otto, Terri Norris, Alex Wild, Aaron Harmer, Zheng Yi Liu, Fiona Cross, M Jithesh Pai, Niels De Rocker, Frank Hecker, Daniel Williams, and Seidai Nagashima for their photos. Finally, I have dedicated this book to Robert Jackson. Without him, I would not have embarked on the career that I have. Thank you.

Ximena Nelson

PICTURE CREDITS

2 Jeremy Squire, 3 alslutsky/Shutterstock, 4 TL alslutsky/Shutterstock, TR Eric Isselee/Shutterstock, MR & BR Chase D'animulls/Shutterstock, 5 TL alslutsky/Shutterstock, TR Protasov AN/Shutterstock BL Chase D'animulls/Shutterstock, BR Lauren Suryanata/Shutterstock, 6 tea maeklong/Shutterstock, 7 Jeremy Squire, 8 serpeblu/Shutterstock, 9 Lou Staunton, 10-11 common human/Shutterstock, 15 & 16 Jeremy Squire, 17 Lou Staunton, 18 & 19 Jeremy Squire, 20-21 Kim Taylor/Nature PL /Alamy Stock Photo, 22 Klaus Kaulitzki/Shutterstock, 23 top ©Matjaž Kuntner, bottom STEVE GSCHMEISSNER/SCIENCE PHOTO LIBRARY, 25 Lou Staunton, 27 Jeremy Squire, 28 Ozgur Coskun/Shutterstock, 29 Jeremy Squire, 30 left D. Kucharski K. Kucharska/Shutterstock, right Neil Bromhall/Shutterstock, 31 Matthew L Niemiller/Shutterstock, 33 top Frank Deschandol & Philippe Sabine/BIOSPHOTO/Alamy Stock Photo, 33 Ingo Arndt/Nature Picture Library, 36-37 Lou Staunton, 38 & 39 Jeremy Squire, 40 Lou Staunton, 41 Jeremy Squire, 42 Martin Pelanek/Shutterstock, 43 Simon D. Pollard, 44 PHOTO FUN/Shutterstock, 45 Fotos593/Shutterstock, 46 Scenics & Science/Alamy Stock Photo, 47 Pong Wira/Shutterstock, 48 nechaevkon/Shutterstock, 49 Ondrej Michalek/Shutterstock, 51 Protasov AN/Shutterstock, 52 Jeremy Squire, 55 Scenics & Science/Alamy Stock Photo, 57 Sabena Jane Blackbird/Alamy Stock Photo, 59 Marshal Hedin, 61 Jurgen Otto, 63 Jeremy Squire, 65 Federico Crovetto/Shutterstock, 68 Michael Hutchinson/Nature PL/Alamy Stock Photo, 69 left Stephen Dalton/Nature PL, right isf photography/Alamy Stock Photo, 71 PREMAPHO-TOS/Nature PL, 72 Bernard Castelein/Nature PL, 73 Jeremy Squire, 74 top Jeremy Squire, bottom Roger Coulam /Alamy Stock Photo, 75 Jeremy Squire, 76 inga spence /Alamy Stock Photo, 77 Martin Harvey/Avalon.red/Alamy Stock Photo, 78 Paul Rollins/Alamy Stock Photo, 79 Kit Leong/Shutterstock, 80 Terri Norris, 82 SDym Photography/Alamy Stock Photo, 83 Lenu/Shutterstock, 85 Gael Rossi/Shutterstock, 87 Olaf Leillinger, CC BY-SA 2.5 <https://creativecommons.org/licenses/by-sa/2.5>, via Wikimedia Commons, 89 ©Alex Wild/alexanderwild.com, 91 Aaron Harmer, 93 Terri Norris, 95 Lou Staunton, 99 guraydere/Shutterstock, 100 RudiSteenkamp, CC BY-SA 4.0 <https://creativecommons.org/licenses/by-sa/4.0>, via Wikimedia Commons, 101 Roland Seitre/Nature PL, 102 blickwinkel/G. Kunz/Alamy Photo, 104 left ©Alex Wild/alexanderwild.com, right Lou Staunton, 105 A_Lesik/Shutterstock, 106 Crystite licenced/Alamy Stock Photo, 107 & 108-109 Jeremy Squire, 110 Pictorial Press Ltd/Alamy Stock Photo, 111 mark phillips/Alamy Stock Photo, 112 IUCN. 2022. The IUCN Red List of Threatened Species. Version 2022-2. www.iucnredlist.org. Accessed September 2023, 113 top Pavel Krasensky/Shutterstock, bottom Rudmer Zwerver/Shutterstock, 114 Matteo photos/Shutterstock, 115 DR P. MARAZZI/SCIENCE PHOTO LIBRARY, 116 Keith Skingle/Alamy Stock Photo, 117 Joe Blossom/Alamy Stock Photo, 118 Albert Wright/iStockphoto, 119 Milan Zygmunt/Shutterstock, 121 Emanuele Biggi/FLPA/Agefotostock, 123 Auscape\UIG/Agefotostock, 125 Heiko Bellmann/Frank Hecker Naturfotografie, 127 Pong Wira/Shutterstock, 129 Zheng Yi Liu, 133 Morley Read/Alamy Stock Photo, 134 left Jeremy Squire, middle Korovin Aleksandr/Shutterstock, right sircco/iStockphoto, 136 & 137 Lou Staunton, 139 D. Kucharski K. Kucharska/Shutterstock, 140 Lou Staunton, 141 Simon D. Pollard, 142 Custom Life Science Images/Alamy Stock Photo, 143 top Milan Zygmunt/Shutterstock, bottom Premaphotos/Alamy Stock Photo, 144 Lou Staunton, 146-147 Emanuele Biggi/Nature Picture Library, 148 rsooll/Shutterstock, 149 Edy Pamungkas/Shutterstock, 153 Wirestock, Inc. /Alamy Stock Photo, 155 Fiona Cross, 157 Ernie Cooper/Shutterstock, 159 Gerhard Koertner/Avalon.red/Alamy Stock Photo, 161 M Jithesh Pai/@wildphotostories_by_jithesh (Instagram) 163 Judy Gallagher, CC BY 2.0 <https://creativecommons.org/licenses/by/2.0>, via Wikimedia Commons, 166 JEWnnn_YEEpld/Shutterstock, 167 top Rudmer Zwerver/Shutterstock, bottom blickwinkel/McPHOTO/M. Schaef/Alamy Stock Photo, 168 & 169 MELVYN YEO/SCIENCE PHOTO LIBRARY, 170 & 171 Jeremy Squire, 172 Vita Serendipity/Shutterstock, 173 top Federico.Crovetto/Shutterstock, bottom adrian hepworth/Alamy Stock Photo, 174-175 Ondrej Michalek/Shutterstock, 176 Morley Read/Alamy Stock Photo, 177 Seidai Nagashima, 178-179 M Jithesh Pai/@wildphotostories_by_jithesh (Instagram), 180 Stephen Dalton/Avalon.red/Alamy Stock Photo, 181 left Michael & Patricia Fogden/Nature Picture Library, right Janice Chen/Shutterstock, 182 shellhawker/iStockphoto, 183 Gerry Pearce/imageBROKER.com GmbH & Co. KG/Alamy Stock Photo, 184 Jeremy Squire, 185 MELVYN YEO / SCIENCE PHOTO LIBRARY, 187 top Paul Bertner/Nature Picture Library, bottom Rachel Miller/iStockphoto, 189 Alf Jacob Nilsen/Alamy Stock Photo, 191 Chris Moody/Shutterstock, 193 nickdor/iStockphoto, 195 (c) Reynante Martinez, some rights reserved (CC BY), Attribution 4.0 International (CC BY 4.0), 197 Andy Newman's Tarantulas/Alamy Stock Photo, 199 Daniel Williams/@macrotheory, 203 M Jithesh Pai/@wildphotostories_by_jithesh (Instagram), 204 Northwest Wild Images /Alamy Stock Photo, 206 & 207 Jeremy Squire, 207 M Jithesh Pai/@wildphotostories_by_jithesh (Instagram) 208 Jane Rix/Shutterstock, 209 zaidi razak/Shutterstock, 211 Vinicius R. Souza/Shutterstock, 212 Cyril Ruoso/Nature Picture Library, 213 Clark Warren/iStockphoto, 216 Stephen Dalton/Avalon.red/Alamy Stock Photo, 218 Image by Ximena Nelson, 220 samray/Shutterstock, 221 Bryan Reynolds /Alamy Stock Photo, 223 Andre Zambolli/Shutterstock, 225 Pascale Gueret/Shutterstock, 227 JAH/iStockphoto, 229 Niels De Rocker, 231 Jeremy Squire, 233 Premaphotos/Alamy Stock Photo, 235 Gucio_55/Shutterstock, 238-239 KirsanovV/iStockphoto, 240 Jeremy Squire, 241 Simon D.Pollard, 242 & 243 Jeremy Squire, 244-245 Agung Y.P/Shutterstock, 246 Colin Marshall/Biosphoto/Alamy Stock Photo, 247 & 248 ©Matjaž Kuntner, 249 Nature Picture Library/Alamy Stock Photo, 250 PREMAPHOTOS/Nature Picture Library, 251 Marek Mierzejewski/Shutterstock, 253 JOHN SERRAO/SCIENCE PHOTO LIBRARY, 254 Alex Hyde/Nature Picture Library, 256 Eric Isselee/Shutterstock, 257 Steve Shinn /Alamy Stock Photo, 258 Alex Hyde/Nature Picture Library, 260 Alex Hyde/Nature Picture Library, 261 Vinicius R. Souza/Shutterstock, 262 Tony Heald/Nature Picture Library/Alamy, 264 PREMAPHOTOS/Nature Picture Library, 265 M Jithesh Pai/@wildphotostories_by_jithesh (Instagram), 267 Jan Hejda/Shutterstock, 269 Lou Staunton, 271 G.Wolf/Zoonar/Alamy Stock Photo, 273 Marshall Hedin, 275 Terri Norris, 277 Jurgen Otto, 278-279 ©Alex Wild/alexanderwild.com, 282 Olhastock/Shutterstock, 283 Kurit afshen/Shutterstock